CHICKENS

THEIR NATURAL AND
UNNATURAL HISTORIES

CHICKENS

THEIR NATURAL AND UNNATURAL HISTORIES

Janet Lembke

Skyhorse Publishing

Skyhorse Publishing books may be purchased in bulk at special discounts for sales promotion, corporate gifts, fund-raising, or educational purposes. Special editions can also be created to specifications. For details, contact the Special Sales Department, Skyhorse Publishing, 307 West 36th Street, 11th Floor, New York, NY 10018 or info@skyhorsepublishing.com.

Skyhorse Publishing® is a registered trademark of Skyhorse Publishing, Inc.®, a Delaware corporation.

Visit our website at www.skyhorsepublishing.com.

10 9 8 7 6 5 4 3 2 1

Library of Congress Cataloging-in-Publication Data

Lembke, Janet.
    Chickens : their natural and unnatural histories / Janet Lembke.
        p. cm.
    Includes bibliographical references.
    ISBN 978-1-62087-055-6 (hardcover : alk. paper)
1. Chickens—History. I. Title.
    SF487.7.L46 2012
    636.5—dc23

PERMISSIONS
Gary Whitehead has graciously given permission for the use of his poem "A Glossary of Chickens," which first appeared in *The New Yorker*, May 24, 2010.

Printed in China

Contents

Acknowledgments....................................vii

1. Chicken Dreams.................................. 1

2. The Ur-Chicken................................. 8

3. The Classical Chicken........................... 18

4. The Medieval Chicken.......................... 35

5. The Renaissance Chicken....................... 39

6. The Medicinal Chicken 48

7. The Transitional Chicken....................... 56

8. The Modern Chicken 65

9. The Resurrected Chicken....................... 85

10. The Conquering Chicken 94

11. Eggs... 103

12. The Scientific Chicken........................ 115

13. The Storied Chicken 131

14. Chicken People................................ 148

15. Chicken Cuisine 166

16. Hen Music.................................... 193

    Appendix: The Blessing of the Hens..... 211

    Notes... 214

    Bibliography 220

Acknowledgments

Nancy Carter Crump, food historian, for her culinary enthusiasm.

Jerry Dodgson, who adopted RJF #256 and told me her history.

Yasushi Ito, of Kobe, for helping research the Japanese chicken.

Boria Sax for sharing his PowerPoint presentation "The Fighting Cock and the Brooding Hen: Chickens as Symbols of Gender in Folklore and Literature."

The chicken people who all glowed with enthusiasm.

The fourteen hens that have brought me much joy.

Lilly Golden, my editor, for her continuing encouragement.

CHAPTER ONE

Chicken Dreams

Chickens! Years after they had begun, my sleepy chicken dreams took feathered form in an unexpected, almost explosive way. Fourteen hens in my backyard! What had I been thinking?

I had been thinking that it would be fun to keep a few birds—hens only, for roosters crow whenever they feel an unstoppable urge to crow, be it at dawn, dusk, or midnight. I live in Staunton, Virginia, a small town smack-dab in the middle of the mountain-guarded Shenandoah Valley. My neighborhood, located in one of the town's five historic districts, features small yards and big houses set fairly close together. Round-the-clock cockadoodledooing would hardly be fair, not just to the people next door but also to those living on several streets nearby. Only hens, sweet hens, would live in my yard.

Hindsight informs me that the chicken-keeping notion began subtly when I visited my veterinarian daughter, Lisa, two decades ago. Two cages sat on the island in her kitchen, each inhabited by a pullet just sprouting her feathers. With those quills bristling from their bodies, they looked like avian versions of the porcupine. During every waking moment, the two held soft, cooing, peeping, chirping, musical conversations. Lisa also kept an outdoor rooster and hens. The rooster was a monster in both size and temperament. Part of his problem was that he was a bird bred for breast size and thus for slaughter, an event that usually takes place when a poult is only six weeks old. This rooster, however, had been allowed to grow up, and his chest was so large that he had trouble standing upright. He was as big as an eighteen-

⌃ Two little ones left behind

pound tom turkey. The rest of his problem was simply that he was a rooster, doomed to be mean and territorial. For him the grim reaper took the form of Biscuit, a rescued greyhound, who thought that chickens were better than rabbits as objects of the chase.

But I dreamed only of hens like Lisa's softly murmurous birds. Her two were not big-breasted creatures meant to end up on Styrofoam trays wrapped in plastic in a supermarket meat department. They were leftovers from a seminar in which she had demonstrated to owners of fancy show birds how to draw their blood to test for pullorum disease, a highly contagious diarrheal salmonellosis.

"The seminar was over," Lisa said. "Everybody else had gone, taking their chicks with them. But there were these two little ones left behind, *peep-peeping* away. So I said *peep-peep* right back, and here they came. It was freezing out; so, I wrapped them up in that old blue throw I keep in the car and took them home."

But the pullets outgrew their cages and were put in the fenced chicken yard to join Lisa's existing flock. However much speed and agility Lisa's pullets had, fences did not deter Biscuit from her chosen sport. Like a hawk amid sparrows, she slaughtered all of them (unlike a hawk, she did not dine upon them). Since then, Lisa has kept many animals, mainly wheezing cats and deformed dogs. But chickens have not been part of her equation. Nonetheless, when chicken-keeping moved to the forefront of my bucket list several years ago, she sent me a book on the care and feeding of chickenkind.

My thoughts about a backyard flock were encouraged by several stimuli. One was an arm's-length consideration inspired partly by "The It Bird," an article by Susan Orlean that saw 2009 publication in *The New Yorker*. The chicken has become an object of desire for lots of city folks. As the February 2011 issue of *Sunset* magazine says, "The chicken coop is the new doghouse. As the backyard chicken craze spreads like wildfire, Fidos . . . are having to share the yard with the ladies." I've also heard it said that the chicken-sitter is the new dog-walker because of the upswing in chicken popularity. The reasons for the upswing are manifold, from enjoying the friendly silliness of chickens to a new emphasis on the locavorian growing of one's own food, be it vegetables or meat or eggs. Through the sale of their eggs, chickens can more than pay for their feed and the straw or pine-bark bedding in the coop. More important, backyard chickens enjoy happy, cage-free lives and gobble up bugs, which enhance the color and taste of their eggs. In the matter of habitats, Ms. Orlean extolled the Eglu, which looks to me like a chicken-sized plastic cave that somewhat resembles an open eggshell. She purchased a model that included two hens in the purchase price. The Murray McMurray Hatchery, which was founded in 1917, supplied the birds.

Next steps: Order the McMurray catalog and check on Eglu pricing. An Eglu with a steel-mesh run that holds four hens or six bantams costs close to a thousand dollars when shipping charges are factored in. Granted, it comes in an enticing array of colors—bold red, hot pink, sizzling orange, sky blue, and soft green—but a thousand dollars amounts to a considerable expense, especially for someone interested in eggs and friendship. McMurray's catalog is a beautifully illustrated encyclopedia of chicken breeds, from plump and plain-looking white, black, or red birds to birds with frizzled feathers or crests or necks completely bereft of feathers. Nor does the catalog stop at chickens but includes turkeys, ducks, geese, pheasants, quail, partridge, guinea fowl, and peacocks. For people inclined toward the exotic, black and trumpeter swans are available, along with demoiselle cranes. Of course, McMurray sells every

imaginable sort of accessory for its birds—brooders, drinking fountains, electric fencing, and more, more, more. The standard order for birds involves getting twenty-five day-old chicks. What would I do with twenty-five peeping babies that needed not only warmth but also protection from predators?

«   A sunny yellow Eglu

Inside, my two cats, hard-wired to respond to motion, would be fascinated by such lively new toys. Outside, possums and skunks roam; both relish chickens and their eggs. The Eglu and the catalog shoved my chicken dreams into a remote corner of my imagination.

But other nudges came along. Because humankind cannot eat grass, most of my yard, front and back, has been turned into a garden. There, a year's worth of tomatoes, green beans, limas, peas, broccoli, winter squash, and garlic grow, along with jalapeño and Serrano peppers for pickling. What would a homemade veggie delight sandwich be without pickled peppers? I also plant seasonal vegetables like carrots, radishes, lettuce, and eggplant. The soil in my yard is basically rusty-red clay studded with nuggets of limestone, some like marbles, others as big as bricks. Clay's virtue is that it retains water. Its faults, though, are many. Earthworms do not find it easy to tunnel through clay, nor can carrots grow straight in such inelastic stuff, especially when it's full of impediments. So, I've built raised beds and added many amendments. The backyard has benefited from two loads of topsoil brought in by dump truck after my husband, the Chief, had demolished the cistern that had provided house and yard with water before the city built mains. The cistern, constructed of bricks, yawned like a giant pit, ready to trap an unwary dog, cat, or child. It had to go. The Chief rented a Bobcat, knocked the bricks into the hole, brought in the topsoil, and spent a grand day roaring around the yard to spread it. The soil over the cistern sank a bit; it has provided a perfect place for a circular raised bed that has since supplied me with strawberries, butternut squash, and lima beans.

The front yard, also red clay, was not so easily and quickly amended. Too small for a Bobcat and a truckload of topsoil, it had to be treated by hand to attain its current loamy state. And what a variety of things it has received! I fed it compost from the bin in the backyard, worm castings from the local farmers' market, alfalfa meal bought in a fifty-pound bag to dress a one-time onion patch, and rabbit pellets donated by a friend, who kept a lop-eared bunny named Alex.

Then, marvelous to say, in a gardening class, I met Geri Maloney who, with her husband, raises chickens for Tyson. "Geri," I asked one evening, "do you have any chicken litter to spare?"

"We don't always," she said, "but right now we do."

Memories surfaced of watching plays in an outdoor theater located near an egg farm. When evening breezes blew the wrong way, *pee-yew!* "Does it reek to high heaven?" I asked.

"Some of it's been sitting and curing a while. Only smells bad when it's wet. That's what you want—composted poop and straw."

We made a date. I borrowed a little white Toyota pickup, received enough litter to fill its bed, and drove home. Geri had warned me that moving the stuff around could stir up fine dust. So, with a mask over nose and mouth, I spent the rest of the day shoveling shit. Well-composted indeed, it had very little odor but did contain lots of feathers and a few complete birds, none of them bigger than sparrows, that had evidently been bogged down in the muck.

The litter was spread over front garden and back. The results were nothing short of miraculous. The Better Boy tomatoes grew as plump as Beefsteaks; the Red Pontiac potatoes, usually about as big as tennis balls, attained football size. Clearly, chicken litter inspired these growth spurts—and acted as an incentive for me to take dreams of chicken-keeping from the back of my mind and bring them closer to the fore. But when litter was needed, Geri could provide. No need for a backyard flock.

Just the same, hither and yon within the city limits, one could hear roosters. Sometimes, a hen or two could be seen scurrying in somebody's garden. Then one day, Frank Strassler, the director of the Historic Staunton Foundation, an organization that oversees historic preservation in a town close to three centuries old, said, "You know, you can have chickens within the city limits."

"Wow! Really?"

"Yes."

But, skeptical, I called the city government's information line. "I've heard that chickens are allowed in Staunton. Is that so?"

The woman on the other end of the phone said, "Let me check." After at least a minute of silence, she returned to announce in stentorian tones, "No fowl!"

There were certainly fowl in town—but, as it seemed, fowl with no legal status. When next I saw Frank Strassler, I said, "City Hall tells me absolutely no chickens."

What a temptation, though, to join the ranks of illegal fowl-keepers! It would not be the first time that I'd committed civil disobedience. That occurred in 2008 when I committed civil disobedience by repainting the yellow lines on either side of my driveway. The town had refused to renew the paint because to do so would violate the rules governing an historic district. But my fresh yellow lines kept people from parking across the drive. Most satisfactory!

As it turned out, though, I'd be on firm ground with chickens. After Frank learned that City Hall had put the kibosh on chicken-keeping, he said, "They haven't read their rules lately." He printed the pertinent section of Staunton's city code, highlighted the parts that deal in particular with poultry, and gave me the seven-page printout for Chapter 6.05.

The code states that only the keeping of sheep and pigs is forbidden within the city limits. As for chickens, section 6.05.060 deals with "Fowl or livestock running at large." It reads: "It shall be unlawful for any person to allow or permit any fowl, poultry, or livestock of any description to stray to public property or private premises of another, and all poultry, fowl, and livestock shall be sufficiently housed or fenced by the owner or person exercising control over the same, so as to prevent the same from trespassing or straying." Nowhere does Chapter 6.05 limit the number of chickens. Nowhere does it prohibit roosters.

But what woman with her wits about her wants her chickens crossing the road? A coop is a given. And I have a yard with gates and chain-link fencing.

Frank did issue a caveat. "I was in a chicken co-op with some friends. The idea was getting our own eggs. But we gave the birds organic feed. Expensive! A dozen eggs cost us at least twice as much as they do in the store."

My chickens wouldn't get organic feed. They could subsist on ordinary laying mash, amply supplemented by the kitchen scraps, bugs, and worms in my compost bins. They could graze on the little grass of the bit of lawn that remains in the yard. They would, however, be denied scratching for food in my raised beds, for the beds would be covered with pop-up tents of green netting. My flock was coming ever closer to cooing, clucking reality.

Then a hitch developed. "Would you like to have my rabbit?" a friend asked. "For free along with her hutch?"

I'd met the rabbit in question—a small Holland Lop with a brown-and-gray marbled coat. "I suppose. Rabbits are pretty good poop factories. Why give her away?"

"She's started to bite the hand that feeds her."

Experience had taught me that gardens relish rabbit poop, and it has an advantage over chicken litter: It need not be composted but can be applied immediately as fertilizer. "I'll take her," I said.

She'd been bought as a pet for a two-year-old, but rabbits make unwieldy pets for someone so young because they can—and do—kick like mules. Rabbits can be delightful pets, but to achieve that state, they need to be played with. This little animal had never been socialized. Living in her hutch, her only contact with humankind had been the hand that fed her. Her

⌃ Holland Lop rabbit

hutch was a hexagonal palace made of wood, with a shingled roof and a white rabbit weathervane. It had three stories, an upper sleeping area, a little shelf midway, and a large lower area for bowl and chew-toys. I don't know what her people had called her. To me she was simply Bunny. She did not bite me, but she flinched whenever my gloved hand touched her. She did bite several people unwise enough to try to pat her. I posted a sign on one of her palace's walls:

LAPIN BIZARRE
DO NOT PUT FINGER
IN HUTCH.
SHE BITES.

Chicken dreams went into abeyance. A rabbit would be easier to care for.

CHAPTER TWO

The Ur-Chicken

In the order of things, chickens are classified this way: Kingdom, Animalia; Phylum, Chordata; Class, Aves; Order, Galliformes; Family, Phasianidae; Genus, Gallus; Species, gallus; Subspecies, domesticus. Translated, these terms amount to Animal with Backbone in the Bird class and the order of Chicken-shapes, which belong to the Pheasant family and are generically and specifically the Domestic Chicken Chicken. But before human beings thought up this order of things, there were chickens. And they didn't care two cackles about what they were.

Birds that we would recognize as chicken-like appeared in the early Oligocene. Thus, they've been running around, scratching through leaves, and eating bugs for some 35 million years. In their beginnings, early fowl, including the ur-chickens, ur-turkeys, ur-geese, and ur-ducks, clucked, gobbled, and squawked before evolution ever produced hawks and eagles (perhaps so that when the raptors arrived, they would find their dinners waiting). And all this took place thousands of millennia before the ancestors of *Homo sapiens* rose up on their hind legs and walked.

But where did the forerunners of modern-day poultry come from? They came from those thunder-lizards, the dinosaurs—that's where. Nineteenth-century scientists posited the descent of birds from dinosaurs. Notable among them was the physiologist and anatomist Thomas Henry Huxley, a disciple of Charles Darwin. This thesis found reinforcement in one of the premier discoveries of the late twentieth century: All avian species, from chickens and chickadees to woodpeckers and wrens, did indeed evolve from dinosaurs, and not just any dinosaurs but those known as theropods, "beast-footed" creatures, which walked on their hind legs (their forelegs were tiny in comparison) and dined mostly on flesh, though some were dedicated vegans. One of the best-known theropods is *Tyranno-saurus rex*, a reptile of truly intimidating proportions. (Imagine a chicken *that* big.)

Many of the theropod dinosaurs, however, were smaller agile creatures that (to the joy of bird-watchers and chicken-keepers) survived the massive Cretaceous extinction—the K/T Extinction Event—that occurred at the end of the Mesozoic Era and did in most other dinosaurs. These theropods darted upright through the swamps and jungles of the Creta-ceous Period about 65 million years ago. Fossils show that a group of them, the coeluro-saurs—the "hollow-tail lizards"—had features remarkably like those that would characterize the birds of the future. One kind of coelurosaur, the maniraptor or "hand-seizer," is likely to have been one of the great-great-granddaddies of birds. A lovely feathered specimen that probably behaved like a roadrunner is *Rahona ostromi* or "Ostrom's cloud," which was found in Madagascar in 1995. *Ostromi* honors John Ostrom, a Yale scientist who catalogued the points of similarity between maniraptors and birds. Those points are wondrously many. A maniraptor's bones were thin and hollow. Its hands had claws, and its eyes were big and round. Its collarbone was fused to form a wishbone. Its foot had four toes, three of which faced forward to support the animal. It had gizzard stones in its digestive system to help grind food. It laid and brooded eggs; science has since discovered that the egg-laying anatomy and the microstructure of eggshells is similar in both maniraptor and bird. It had scales, produced by thick-ened patches, called placodes, on its epidermis. Because

Rahona ostromi, a birdlike dinosaur

scale-forming placodes deposit new cells on just one of their sides, scales are horizontal. Somewhere along the evolutionary line, the placodes underwent a genetic mutation that caused them to deposit new cells in a cylindrical ring, creating vertical filaments and bristles. These proto-feathers insulated theropod legs and bodies and thus helped to maintain body temperature. It may well be that these rudimentary feathers were used in sexual displays. A science writer who has studied the evolution of feathers notes that alligators and crocodiles are the closest living relatives of birds and dinosaurs. He writes, "Although these scaly beasts obviously do not have feathers today, the discovery of the same gene in alligators that is involved in building feathers in birds suggests that perhaps their ancestors did, 250 million years ago before the lineages diverged. So perhaps the question to ask, say some scientists, is not how birds got their feathers, but how alligators lost theirs."

Like its bird descendants, the maniraptor had no urinary bladder. Egg-laying and the absence of a bladder need a few words. Imagine trying to become airborne in the presence of internal weights that insist on obeying gravity. External reproduction helped lift flying creatures aloft because they did not have to carry their embryonic young. And, as we know from observing bird-splat on our windows and cars, the stuff that passes for bird urine has considerable substance. Most of its water has been resorbed, thus eliminating another weight.

R. ostromi, however birdlike it looked in its encasing feathers, was not a bird but rather a dinosaur. Its classification goes this way: Kingdom, Animalia; Phylum, Chordata; Class, Reptilia; Superorder, Dinosauria. In other words, it was an animal with a backbone that belonged to the reptiles and, in particular, the dinosaurs.

Non-dinosaurian birdkind actually got off the ground some 161 million years ago in the Late Jurassic period. Scientists have speculated from fossil evidence that nature may well have experimented early on with a four-winged version. The ur-birds, however many wings they had, coexisted with pterosaurs, "winged lizards," which were beaked, birdlike reptiles probably capable of short bursts of powered flight. Pterosaurs represent, however, an example of convergent evolution—in this case, the development of similar but not related flying creatures. They died, along with most other kinds of dinosaur, in the K/T Extinction Event. The best-known proto-bird of the Jurassic was *Archaeopteryx lithographica*, or "ancient feather drawn in stone." Its first fossil, found back in 1860, was that of a single feather, and it came

≽ Archaeopteryx

from a limestone deposit in Bavaria, where subsequent fossils, some of them whole skeletons, have since come to light. This particular ur-bird exhibits theropod characteristics and seems likely to have had theropod ancestors. Like them, it had teeth, there were claws on the ends of its feathered wings, and its backbone stretched out into a tail. About the size of a big crow, it could fly but only in a clumsy, short-distance fashion (much in the manner of, oh yes, a chicken). Science classifies it as a true bird: Kingdom, Animalia; Phylum, Chordata; Class, Aves. *Aves* is the Latin word for "birds."

Another equally ancient member of the Aves, the true birds, was *Confuciusornis sanctus*, or "holy Confucius-bird." It was discovered in China in 1994. Hundreds more have since been found. About the size of a pigeon, it looked a bit more like a modern bird with its horny, toothless beak, and short, fused tailbones. But, like archaeopteryx, it could not lift its wings above its shoulders, and thus was not capable of sustained, wing-flapping flight. Ancient Feather and Confucius-Bird are not to be thought of as the direct progenitors of today's more than nine thousand avian species. Despite their inclusion in the Class Aves, they were primitive cousins.

Nor were some of the lineages that evolved during the Cretaceous directly ancestral,

≽ Enanthiornithes, a true bird

although they were classed as Aves. Prominent among them were the Hesperornithiformes, the "western bird-forms," that swam (think penguins); the toothed Ichthyornithiformes, the "fishy bird-forms," that most likely fed on fish; and the Enantiornithes, the "opposite birds," that readily took to the air. Reconstruction of one species shows a beautiful creature with a stout finch-like beak and dense, pinkish feathers speckled with brown on head and chest. The shape of the beak is accurate because it was preserved in stone, as were the feathers, though their colors strike me as wishful thinking. No one knows just why Cyril Walker, the scientist who described the birds, called them "opposite," but a bit of guesswork by other scientists posits that "opposite" reflects the fact that they show anatomical features that are inversions of those found in modern birds. These contrarians may represent a convergent evolution in birdkind. Science classifies them as Aves, true birds of the Kingdom Animalia and the Phylum Chordata. But they, the fishy-forms, and the western-forms all suffered extinction in the K/T Event. Chickens and all other modern avian species evolved from the theropod-descended birds that miraculously made it through the massive die-off that killed most other reptiles. Yes, birds, which developed from a long line of reptiles, can rightly be called reptiles. But, because that word summons snakes to the human imagination, we've glommed onto Aves, a word that nicely camouflages reality. Luckily, no birds are venomous. Would we find snakes more pleasing if they had feathers?

The trick that sets many birds apart from their ancestors is flight. Granted, not all birds take to the air. The ratites, like the emu, ostrich, and cassowary, are earthbound, as is the completely wingless kiwi and the penguins, which have flippers. Theories have it that airborne birds are descended either from arboreal theropods that gradually evolved wings to glide, then fly or, more likely, from terrestrial theropods that achieved flight in three stages. First, mini-wings that looked like tissue stretched under forelimbs acted as parachutes; second, when larger wings developed, flapping them led to flight; and, third, fully formed wings sent birds aloft to soar. But most Galliformes are not high fliers; they're terrestrial birds, with most of them given to only short bursts of flight. And, as we all know, domestic chickens have a hard time getting off the ground, and when they do, they don't stay up there very long.

Two questions regarding chickens pop up persistently: Why did the chicken cross the road? And, which came first, the chicken or the egg? The second question falls into a category known as a causality dilemma—which of two possibilities was the prime cause? The answer to this one is that because the dinosaur came first and because it laid eggs, the egg antedates the chicken by millions of years. If you rephrase the question to pose a new causality dilemma—Which came first, the dinosaur or the egg?—a scientific answer can be provided. It depends upon the zygote, which is the first cell created by the union of male sperm and female ovum. And the zygote is the place in which small mutations in DNA occur. Here's what the website How Stuff Works has to say: "Prior to that first true chicken zygote, there were only non-chickens. The zygote cell is the only place where DNA mutations could produce a new animal, and the zygote cell is housed in the chicken's egg. So, the egg must have come first." This explanation needs slight alteration, for it presupposes a mutant zygote in a chicken's egg. It should read simply that the zygote is housed in the egg.

As for the first question, let's get that one out of the way right now before this exploration of our earliest chickens moves on to less frivolous realms. Conventional wisdom has it that the chicken crossed the road to get to the other side, but given the bird's tendency to one-upmanship among its own kind, it crossed the road to show the possum that it could be done.

A question of considerably more interest is: When and where did our domestic chickens originate? The answer is: in the forests of northeastern India, southern China, Malaysia, and other parts of southeastern Asia. An intermediary between yesterday's birds and today's chickens was the Malaysian russet-colored megapode, a mound-builder. And this megapode—the word means "big foot"—was transformed in the passage of time to the immediate

⤳ Red jungle fowl cock and hen

wild ancestor of chickens, a wild ancestor that is still very much in evidence in its original home. It's the red jungle fowl, strutting and scratching through leaf litter. Science calls it *Gallus gallus gallus*, "Cock cock cock," a name that comes close to crowing in its redundancy. The third *gallus* denotes its subspecies, for it is the primary member of a flock of six subspecies, found generally in Asia, except for one—*G. g. domesticus*, the "domesticated cock cock," or our everyday chicken, which has certainly conquered the world.

Several species of jungle fowl apart from the red may also be found today in Asia: the grey jungle fowl, *G. sonneratii* or "Sonnerat's cock," of

India (Pierre Sonnerat, 1748–1814, was a French naturalist and explorer); the green jungle fowl, *G. varius* or "vari-colored cock," of Java; and the Sri Lankan jungle fowl, *G. lafayetii*. The hens of all species are uniformly drab birds, half the size of their mates and clad in brownish feathers, but the roosters are glorious, every last one wearing a veritable Joseph's coat of colors. Though in size he's not much bigger than a bantam, the red jungle fowl is a spectacularly handsome bird, with a distinctively large white earlobe, a cape of long golden feathers pendant from his neck to his shoulders, body decked in brown and bright maroon, an arched and plumy tail with iridescent glints of blue, purple, and green, and wicked spurs on the backs of his legs.

I wonder how many of the birdwatchers who travel the world to add new species to their life lists have included jungle fowl as birds that must be seen. Or have they been dismissed simply as chickens? If they want to see these birds in pure form, they need to hurry, for jungle fowl in every country where they are endemic now suffer hybridization due to interbreeding with *G. g. domesticus*. The International Union for Conservation of Nature (IUCN) deems them birds of least concern on its red list of endangered species. After all, a chicken is a chicken is a chicken. Why fret?

Science now declares that our backyard birds are indeed directly descended from red jungle fowl with a dash of Sonnerat's thrown in for good measure. The next question is, why domesticate this wildling? One reason has nothing whatsoever to do with its edibility, although it and its eggs, as well, must have been served up not infrequently for supper. One of the original bargains involved in domesticating the red jungle fowl took place in the Indus Valley of India in the fourth or third millennium BC. It was not one that traded care for eggs, meat, and feathers. Rather, entertainment was at the top of the human agenda. Exchanges of money and goods surely took place, too, either as a purchase price for the birds as merchandise or as bets. The primary interest of the keepers lay in cockfighting, a bloody sport that spread rapidly through Asia, the Near East, and Africa. I've heard it said that a gamecock does not know that he is beaten as long as a spark of life remains.

ARKive, a website devoted to presenting pictures and information about the world's endangered species, offers a rip-tearing video of two red jungle fowl roosters battling to see who can claim title to Cock of the Walk. The two engage in a sort of avian kickboxing, with one, then the other having at it with his legs, claws, and spurs. Feathered balls of fury, they kick and kick again, lifting each other into the air. They tumble, rise, and fight again. The golden feathers in their capes bristle in indignation. It's easy to see how such fierce competition between two handsome, highly colored birds whetted a human bloodlust: If they do this as a matter of course, why not take bets? I think that the most spirited money makers must have been caged so that they couldn't get away.

⌃ Chinese zodiac sign for the rooster

In the last few decades, archaeologists have discovered evidence for an eggs-meat-feathers bargain in China that occurred far earlier than the Indus Valley's fourth-millennium domestication. The eggs-meat-feathers bargain may well have taken place in 6,000 BC when people apparently transported red jungle fowl to dry northern China, far from their wonted forest habitat. The chicken bones that have been found came from birds larger than jungle fowl, though they're not so large as the bones of today's domestic fowl. And not just chicken bones but earthenware models of the bird have been unearthed in China. Excavations there have found remains of domesticated chickens in at least sixteen Neolithic sites, all of which predate the Indus Valley civilizations that specialized in cockfights. Other sites in Asia and—surprise!—in Europe also predate the goings-on in the Indus Valley. The jungle fowl was a well-traveled bird.

The Chinese zodiac, which dates back to at least the fifth century BC, attests to the popularity of the bird, for the rooster is one of its twelve animals. Legend has it that when Buddha prepared to cast off his earthly body, he summoned all the animals. Only twelve came: dog, dragon, goat, horse, monkey, ox, pig, rabbit, rat, rooster, snake, and tiger. Buddha decided to name a year after each of those faithful animals. People born in the year of the rooster (as I was) are said to exhibit some stellar qualities, like resilience and courage, and some that are much less desirable, like conceit, impatience, and bossiness.

Cockfighting and chicken-keeping spread rapidly eastward into the South Pacific from China and westward from the Indus Valley. Chicken bones have been found in Egyptian tombs dating back to the Old Kingdom in the third millennium BC. The birds arrived in Greece no later than 700 BC, and the Greeks acknowledged their origins by calling different breeds Median and Chalcidian, names that refer to places in Persia and Syria respectively. A local breed, the Tanagrian, had also been established just north of Athens. It was black with a bright red comb and wattles and a few white speckles about the beak and the tail. The birds from Tanagra and the Near East were said to be unsuited for laying but fit indeed for combat. Nonetheless, the Tanagrian coloration is still common in Mediterranean countries.

Not long thereafter, various strains of domesticated jungle fowl made their way to Rome and thence into northern Europe. Spanish conquistadors introduced the battling birds into the New World, though there is reliable DNA evidence that the Araucana breed native to Chile is pre-Columbian and crossed the Pacific with Polynesian sailors. This South Pacific morph of the jungle fowl is distinguished by its complete lack of a tail, the prominent tufts of feathers on its ears, and the production of pale blue eggs. An occasional bird will lay eggs of a soft olive green. Interbreeding has produced Araucana hybrids with tails and without tufts, but the laying of blue eggs is still characteristic. I have found no evidence anywhere that, once the Araucana breed had been established, the roosters were sent into mortal combat.

Today, cockfighting has been declared illegal throughout the United States. Louisiana was the last state to outlaw it, but in 2008, banned not only cockfighting but gambling on it. Nonetheless, even though the blood sport is outlawed, gamecocks may easily be found. I've seen working birds on a farm only fifteen miles from Staunton. They

⌃ Tailless Araucana pullets

are gaudy, loudmouthed, and spurred with stiletto-like spikes, made more wicked by the attachment of steel spur covers. Gamecocks, though larger, resemble their wild red ancestor more closely than do any other domestic chickens.

For the ur-chicken in all its guises from theropod through archaeopteryx and enantiornithes to early jungle fowl, our main evidence arises from paleontology and archaeology. We have lots of fossils and some clay models, but written records are absent. Along the way, however, the ur-chicken turned into the classical chicken, about which we find much pictorial—and a bit of verbal—evidence in Egypt and Greece. It was in Rome that chicken-keeping found eloquent commentators.

CHAPTER THREE

The Classical Chicken

Today, we associate ancient Egypt with the Nile, pyramids, and the sphinx, hieroglyphics, animal- and bird-headed deities, and rulers both illustrious and notorious, like Tutankhamun and Cleopatra. As it happened, the Egyptians raised chickens in factory farms. The Chinese also went in for large-scale chicken farming, perhaps as early as the fourth millennium BC. A look at Egyptian practice will serve to illustrate what was done in China. The Egyptian bargain may have been, "We'll care for you so that you feed the villagers building the royal tombs in the Valley of the Kings."

The first representation of a chicken in Egypt appears about 1840 BC in the days of the Middle Kingdom. Four hundred years later, the pharaoh Thutmose III, who reigned from 1479 to 1425 BC, acknowledged tribute paid in chickens by an unidentified place in the Near East: "Four birds of this country. They bear every day." And the tomb of King Tut, who ruled from 1333 to 1323 BC, yielded a painted potsherd depicting the head of a rooster. These paintings point to a lively trade in exotica with other countries; they do not begin to suggest that chickens had become a common agricultural animal in Egypt. Cocks, however, were sacrificed to the god Osiris.

The factory farms seem to have gone into business no later than the reign of Thutmose III. Aristotle (388–322 BC) noted the practice but did not understand just what was taking place. "Eggs are hatched by the incubation of birds," he says, "but they are also hatched spontaneously by being placed among dung in the earth, as is the case in Egypt." Eggs are certainly involved, and dung as well. The practice, however, involves no spontaneity; rather, it is ingeniously planned and executed. In the fourth century BC and probably earlier, the Egyptians built hatching ovens, which were large buildings constructed of thick mud bricks. George Sandys, a British traveler who visited Egypt in the early 1600s, describes these wonders as they had existed for well nigh a thousand years: "Here hatch they egges by artificial heat in infinite numbers." The ovens were placed on either side of an arched entrance tunnel, and they were double, each with a lower and an upper floor, crowned by a conical roof with a vent for releasing smoke. The lower floors were covered with mats "and upon them egges, at least six thousand in an oven." The upper floors were actually mat-covered gratings, on which:

> three inches thick, lyeth the dry and pulverated dung of camels, buffalos, &c. At the higher and further sides of these ovens are trenches of lome [loam], a handful deep, and two handfuls broad. In these they burn of the aforesaid dung, which giveth a smothering heat without visible fire.

Straw was also used as a heat source. So that the eggs would not cook before they could hatch, the fires did not burn constantly but were kept going for an hour in the morning and another in the evening. In addition to the incubating ovens, the building had pens for laying hens, for

« Egyptian
hatching
oven

newly hatched chicks, and (I assume) for inseminating roosters. It also contained living quarters for the person who kept the ovens warm. Aside from receiving room and board, he was paid in extra chickens. Experience—how warm or cool the air felt on the skin—must have taught him how to keep oven temperature steady. He also performed the hen's natural task of turning the eggs. What on earth was done with these thousands on thousands of birds, I don't know, but the very existence of such huge-scale factories points to a big market for roast chicken, stewed chicken, and chicken in a multitude of other edible guises.

The earliest Greek representations of the chicken are pictorial. Painted roosters crow and fight on Corinthian pottery dating back to the last years of the 700s BC. They began showing up in stories and poems in the sixth century BC. Cocks and hens figure in many of Aesop's fables. This teller of tales, born about 620 BC, near the Black Sea, served as a slave on the island of Samos but was later freed. All of his cocks cockadoodle with great gusto, and some are cagy enough to evade the blandishments of a fox. One manages to scare off a lion with the shrillness of his crow. Others, however, succumb to cats, thieves, and eagles. My favorite fable is "A Woman and Her Hen." Here it is in the seventeenth-century translation of both tale and moral by Englishman Roger L'Estrange:

> A good woman had a hen that laid her every day an egg. Now she fancy'd to her selfe, that upon a larger allowance of corn, this hen might be brought in time to lay twice a day. She try'd the experiment, but the hen grew fat upon't, and gave quite over laying.
>
> Moral:
>
> He that has a great deal already, and would have more, will never think he has enough 'till he has all; and that's impossible: wherefore we should set bounds to our desires, and content our selves when we are well, for fear of losing what we have.

The elegiac poet Theognis writes of a night out, "Of my own accord, I will go out at eventide and return at dawn with the crowing of the new-awakened cocks." The poet Pindar (c. 522–443 BC), who composed epinician odes to be sung and danced in honor of victors in the great athletic contests, composed "Olympian XII" to celebrate Ergoteles of Himera, who won the boys' long foot race in 466 BC. Ergoteles, born in Crete, had been forced by civil strife to migrate to Himera, an important Greek city in Sicily, a city that had suffered a massive depopulation and so needed foreign immigrants. Crete rarely sent contestants to the great games. If Ergoteles had stayed in Crete, it's likely that he would never have run in the foot race. Pindar sings:

> Son of Philanor, your honor, like that
> of the gamecock that fights
> in his own yard, would have no fame—
> the crown won by your flying feet
> would shed its leaves and disappear
> had not hostilities deprived you of your home in Knossos.

Several Himeran coins, dating back to the mid-sixth century BC, show roosters.

The Greek philosopher Heraclitus of Ephesus (c. 535–c. 475 BC), famous for saying that you cannot step in the same river twice, left a down-to-earth fragment that deals with chickens: "Pigs wash in mud, chickens in dust."

Gamecocks may have had something to do with the Greek triumph over the Persians at the battle of Salamis in 480. Holding up a pair of the combative birds, Themistocles, the commander-in-chief, is reported to have inspired the troops by saying, in essence, "These do not fight for their country nor for their gods nor for their freedom. They fight because neither would suffer defeat." In commemoration, a cockfight was subsequently staged yearly in Athens as a public spectacle. The fights were staged on a table with raised edges.

⌃ Gamecocks in combat

A scholar who has made a study of Greek and Roman household pets declares that "the young bloods of ancient times took great pleasure in the pleasant pastimes of cock-fighting, quail-fighting, and even crane-fighting. To all intents and purposes, these game birds were to the ancient grandee what falcons and hawks were to medieval lords." Even women kept cocks as pets. One representation shows a nursing mother gazing not at her suckling child but rather at a pair of gamecocks. Reliefs on Greek tombs sometimes depict cockfights. The first domestic chickens, which became abundant in Greece in the mid-500s, came from a long line of battling birds. The bargain was still that we would care for them in exchange for their fulfillment of the human propensity for gambling and relishing bloody battles.

In another kind of bargain—I will care for you so that you may help me honor the gods—Greek chickens were held sacred to some deities. Athena's helmet is embellished with a cock. Eating chickens was forbidden to people who participated in the Eleusinian Mysteries dedicated to the cult of the grain goddess Demeter and her daughter, Persephone. The earth goddess Cybele's priests, who ritually castrated themselves and dressed thereafter in women's clothes, were known as Galli—cocks. Cockfights took place in theaters sacred to Dionysus, god of wine, song and drama, fertility, and the orgiastic. Chickens were also used as sacrifices. Zeus received sacrifices of white cocks. We also have the word of Plato, the philosopher and student of Socrates (428–348 BC). In the *Phaedo*, Socrates's friends, among them Phaedo and Crito, are gathered for their teacher's final hours. He drinks the hemlock, walks till his legs begin to numb, then lies down, covering his face. Toward the end, he removes the covering and speaks: "Crito, I owe a cock to

Aristotle »

Asclepius. Will you please pay the debt?" Crito makes a positive response. Asclepius was the god of medicine and healing, whom Socrates wished to thank for curing him of life. So, it was not only in fights that roosters shed their blood but also in sacrifice. Nor were the remains left to rot on an altar. No indeed, they were roasted and eaten.

Aristotle paid considerable attention to birds in his *History of Animals*, defining them as egg-laying creatures with two feet like those of human beings except that birds bend them backward. (He was mistaken here, for chickens are digitigrade creatures that walk on their toes; the bend in the leg represents an ankle.) Hens and the laying of eggs intrigued him. In the noteworthy 1910 translation of the *History* by naturalist, mathematician, and Greek scholar Sir D'Arcy Wentworth Thompson, Aristotle wrote that:

> Some birds couple and lay at almost any time of year, as for instance, the barn-door hen and the pigeon: the former of these coupling and laying during the entire year, with the exception of the month before and the month after the winter solstice. Some hens, even in the higher breeds, lay a large quantity of eggs before brooding, amounting to as many as sixty; and, by the way, the higher breeds are less prolific than the inferior ones. The Adrian hens are small-sized, but they lay every day; they are cross-tempered, and often kill their chickens; they are of all colours. Some domesticated hens lay twice a day; indeed, instances have been known where hens, after exhibiting extreme fecundity, have died suddenly.

Some interpretation is in order. The Greek that Sir D'Arcy translates as "higher breeds" and "inferior ones" refers, respectively, to birds true to a breed standard and to birds that are nondescript crosses. The Adrian hens were apparently bantams. As for noting that hens may keel over when they lay an inordinate number of eggs, Aristotle succumbed to credulity. He thought, as well, that hens frightened by thunderstorms laid addled eggs and that hens deprived of brooding would sicken. He observed that after being trodden by the cock, hens would shake themselves and scratch in dry straw, tossing it hither and yon. The comb of the barn-door cock caused Aristotle some confusion. Observing that some birds have a crest that sticks straight up and is composed of feathers, he wrote that he didn't understand the crest of the barn-door cock—though it looks like flesh, he can't determine just what, precisely, it is. And Aristotle offers a method of sexing chicks by the shape of the eggs from which they emerge: "Long and pointed eggs are female; those that are round, or more rounded at the narrow end, are male." Not so, as anyone raising chickens can tell you.

It is Aristotle who made the world's first recorded experiments in embryology by opening chicken eggs in successive stages of incubation. Early on, the heart appears as a speck of blood, which beats from its inception. Veins appear. Then the body comes into being, small and white at first. The head is clear to see with its huge, lidded eyes that diminish in proportion as the head and body grow. Aristotle discovered that the chick forms in the egg white and finds nourishment in the yolk. When the egg is ten days old, the complete chick is plain to see, stomach, viscera, and all. It is protected from the surrounding liquids by membranes. In Sir D'Arcy's translation:

> About the twentieth day, if you open the egg and touch the chick, it moves inside and chirps; and it is already coming to be covered with down, when after the twentieth day is past, the chick begins to break the shell. The head is situated over the right leg close to the flank, and the wing is placed over the head.

Membranes like an afterbirth emerge from the shell along with the chick. And inside the chick will be found much of the yolk, enough to keep it fed till it begins to scratch for itself and eat the mash that is offered. (That the yolk nourishes the chick after it hatches is the reason that day-old chicks can be sent through the mail: There's enough food on hand to keep them alive till they reach their destination.) Aristotle makes much of the wind egg. Such eggs, he states, are laid without prior copulation—they are infer-

Wind egg atop normal eggs »

tile, that is—and they are smaller, less tasty, and more liquid than those produced by a mated bird. Their contents never coagulate properly into yellow yolks surrounded by albumen. Though summer is the peak period for wind eggs, they may also be laid in spring, when they're called zephyr eggs because in that season some female birds have been seen facing into the inseminating wind with bills agape. (In hot weather, it's the instinctive habit of birds to do just that to cool off with a sort of avian panting.) He proposed that hen birds of many species allow themselves to be so fertilized by wind—chickens, geese, peahens, partridges, and doves. Centuries later, the Roman Pliny would state that the third egg of a clutch laid by a pigeon is invariably a wind egg. The reason for this classical brooding on wind eggs may be simply that Greek philosopher and Roman natural historian were attempting to explain the enduring fact of infertile eggs. (The term "wind egg" is still used today but denotes a very small egg laid without a yolk; the shell contains only albumen.)

For all his occasionally wrong-headed notions, Aristotle was a superb natural philosopher who examined the world of nature not just right side up but also inside out and upside down.

Aristophanes (c. 450–388 BC) treats us not to science but to comedy in his antic play *The Birds*, in which two Athenian dissidents seek to establish a cloud-borne kingdom of birds to act as intermediaries between men and the gods. The chorus, led by a nightingale once a queen, comprises many species of birds—francolin, kestrel, sparrow, pigeon, and a full flight of others. A chicken is not among them. But the birds, reminiscing sadly about their onetime sway over the world's kingdoms, speak of their compatriot who once ruled Persia. This Persian bird, well-equipped with spurs and a jaunty crown, is none other than a gamecock, resplendent in his many bright and regal colors. To call him the Persian bird is to recognize the place whence he made his way to Greece and kept his battles going, though in a far less kingly sphere.

⌃ Mosaic of a cockfight, The House of the Labyrinth, Pompeii

Before leaving ancient Greece, I should mention that the cock figures among a slew of mythical birds—the *kepphos*, a seabird that snaps at foaming waves; the Pegasus bird that sports a horse's head; winged griffins with eagle-like beaks and the bodies of lions; the Sirens, those sweet-singing, fatal birds with women's heads; and the phallus bird. The last is none other than the cock, depicted as a bird, the feathered body of which is that of rooster, while its neck and head are an erect penis with a wattle of testicles at the point where the neck of a real rooster joins its breast. Athenian vases portray this bird as a cosseted ladies' pet. Not just in English but in many other languages, the word for "cock" designates not just the bird but also the penis.

In Rome, the gamecock figures in ordinary life. Cockfights were an integral part of entertainment, and Romans of any stripe were avid bettors. Early Christians, too, were as enthralled by cockfighting as any pagan. Bas-reliefs of battling birds can be found on some of their sarcophagi. St. Augustine (354–430 AD) was apparently a fan. In *De ordine—On Order—*he sees a cockfight taking place under divine direction:

Once could see the cocks' heads thrust forward, their combs inflated; they were striking vigorously, very cautiously evading each other, and in every motion of these animals unendowed with reason there was nothing ungraceful, since, of course, another higher reason was guiding everything they did, finally the very law of the victor, a proud song and limbs gathered together like a ball as if in the proud scorn of domination.

The cock has a sure place in myth. Once upon a time out of time when Mars was trysting with Venus, the wife of Vulcan, he put Alectryon on guard at the bedroom door. But Alectryon fell asleep. Because of his lapse, he was changed into a cock and given the task ever after to announce sunrise. (The poor boy was doomed from the start: His name is a Greek word for "cock.") A seventeenth-century Italian writer, Ulisse Aldrovandi (whom we'll meet later), finds a more down-to-earth reason for the cock's association with Mars: "The rooster seems to have been sacred to Mars for this reason: because he tries with the greatest zeal to come off victor in a fight, does not wish to suffer servitude to anyone else, and thus fights to the very death." The cock was associated with other deities, as well. He was sacred to Latona, the mother of Apollo and Diana, and was represented as sitting in the left hand of Mercury, the divine messenger of the gods. Then, because the rooster, acting as a sort of hinge between night and day, wakes us with his crowing, he is also sacred to the moon and the sun.

The cock's ability to frighten a lion figures in Latin literature as well as in Greek fables. The poet and philosopher Titus Lucretius Carus (c. 99 BC–c. 55 BC) has a telling passage in *De rerum natura*—*On the Nature of Things*. (The translator is not given credit, but it's the best rendering that I've found.)

> Lo, the raving lions,
> They dare not face and gaze upon the cock
> Who's wont with wings to flap away the night
> From off the stage, and call the beaming moon
> With clarion voice—and lions straightaway thus
> Bethink themselves of flight because, ye see,

> Within the body of the cocks there be
>
> Some certain seeds, which into lions' eyes
>
> Injected, bore into the pupils deep
>
> And yield such piercing pain they can't hold out
>
> Against the cocks, however fierce they be.

But, Lucretius writes, men need not fear a cock's gaze. Either the darts cannot penetrate our eyes, or if they do, they fall right out.

The comb of a cock was used in Rome to brew a potion that would bring on madness. The satirist Juvenal (AD 60–c. 140) wrote that Rome's third emperor, Caligula (AD 12–41), became insane as a result of drinking just such an anti-tonic, which had been given to him by his fourth wife, Milonia Caesonia. Sexual perversity, overspending, and cruelty characterized the last year of his reign, which was terminated by assassination.

The Roman interest in chickens went far beyond breeding and training cocks for mortal combat. They gave great shrift to the raising of domestic chickens for meat and eggs. In his eightieth year, the farmer-statesman Marcus Terentius Varro (116–27 BC) wrote a book on agriculture, *De re rustica—On Rural Subjects*—in which he discusses in detail the things that every farmer should know, including planting crops, grazing flocks and herds, storing fruits and vegetables, and keeping chickens.

He mentions three kinds of fowl from which a chicken-keeper might select: barnyard, wild, and African. The last of these is actually the guinea fowl, a large, loudly garrulous bird without a hint of music in its voice. What he meant by "wild" fowl has not been determined; scholars debate whether they were partridges, feral chickens, or something else. The barnyard birds are chickens that we all could recognize. Varro zeroes in on

⨠ A cock with a fine comb

five points that the farmer should keep in mind: selection and purchase, breeding, eggs, chicks, and the fattening of birds for the table. He identifies three types of barnyard fowl: the hen, the cock, and the capon. He advises that hens be selected from breeds that lay abundant eggs and also have reddish feathers and black wings, large heads, stand-up combs, and full breasts. When it comes to buying a cock, he suggests that the bird chosen have a reddish comb, a beak that's short, wide, and sharp, eyes dark gray to black, red wattles speckled with a little white, neck parti-colored or golden, well-feathered thighs, short lower legs, long claws, and a large plumy tail. Above all, the cock should be *salax*, eager for sex. But beware—don't buy a cock from Tanagra, Media, or Chalcis. Those birds are bred for fighting, not for taking care of hens. Varro recommends a method for castrating a cock that would today be considered brutal: ramming a red-hot iron into the bird's groin and then smearing potter's clay over the wound. The operation makes it easier to fatten the bird, for he loses not just his crow but also his calorie-consuming sex drive. (Today, a cock becomes a capon when his large testicles, tucked within his abdominal cavity on either side of his backbone, are surgically removed.)

Varro is a prodigious source of advice on chickenkind, raising chickens, and fattening chickens. He specifies the dimensions of a coop with perches, nests, windows, and a door by which the *gallinarius*, the poultry-keeper, can enter. Living quarters for the *gallinarius* are attached to the coop, in much the same way that those who tended Egyptian hatching ovens lived right next to their work. Other Varronian words of wisdom include caring for chicks by picking lice from their heads and necks when their feathers begin to sprout and burning stag antlers around chicken coops to keep snakes from entering. His prescriptions for fattening hens sound like an archaic version of factory farming. After the feathers are plucked from their wings and tails, the hens are confined in small cages with a hole for the head and a hole for the tail, the latter so that they don't soil themselves with their own excrement. After incar-

« Pliny the Elder

ceration, the hens are crammed with such goodies as pellets of barley meal, moistened flax seeds, and wheat bread wetted down with mixed wine and water. After twenty to twenty-five days, they will be plump and delectable.

Amid his discussions of plants, animals, minerals, medicines—indeed, the whole world of nature—Pliny the Elder (AD 23–79) also has a good bit to say about chickens. Book X of his *Natural History* is devoted to birds, beginning with an account of the phoenix. Most of his chicken lore is derivative rather than gleaned from personal observation. He has obviously read Aesop's fables, from which he gained the information that a cock's crow will terrify a lion. He reiterates at length the wisdom that he received from Aristotle and Varro, such as listing desirable and undesirable breeds, determining the sex of a chick from the shape of the egg, and the mating of hens with cocks all year round, except for the two months preceding and following the winter solstice. He relays Aristotle's story of the drunkard in Syracuse who put eggs in the earth and sat there drinking until they hatched. Pliny mentions the addling of eggs caused by a thunderstorm but prescribes a remedy—placing an iron nail beneath the straw in the nest. In Pliny's book, the cry of a hawk also makes eggs go bad, but he offers no defense against that bit of bad luck. One of his most amazing statements is that hens indulge in a religious ritual: "After laying an egg they begin to shiver and shake, and purify themselves by circling round, and make use of a straw as a ceremonial rod to cleanse themselves and the eggs."

Pliny vastly admires the cock, which he calls a "night-watchman" skilled in astronomy. The birds mark the passing of time during the day with a crow every three hours and take to their beds at sunset. At sunrise, they wake to flap their wings and crow again and rouse mere

A rooster crowing the hour »

mortals from their sleep. The crowing of a cock was portentous, indeed. It could cause wood to splinter. It could signify victory in battle. For soldiers, it signaled the hour for the changing of the watch. As a meteorologist, the cock was well-regarded; if he crowed loudly and flapped his wings, rain was imminent.

In Rome, the human bargain with chickens also included giving care in exchange for divination. Pliny brings us reportage on those practices as well as giving us facts and fiction about barnyard poultry. Sacred cocks were tended by a *pullarius*, a chicken-keeper, who would offer them food when their services were needed; if they gobbled it up, good fortune was on its way, but if they disdained the tidbits, cackled, flapped their wings, or flew off, bad luck was imminent. The flight of these birds and their manner of pecking for insects and seeds could also be read for omens. Indeed, in Pliny's view, omens from gamecocks ruled both civilians and the military:

> These birds control our officers of state; these order or forbid battle formation, being the auspices of all our victories won all over the world; these hold supreme empire over the empire of the world, being as acceptable to the gods with even their inward parts and vitals as are the costliest victims.

So, auguring yea or nay, chickens ruled the Roman roost.

But the statesman and orator Cicero (106–43 BC) scoffed at the business of letting chickens take the place of common sense. In *De divinatione—On Divination*—he has this to say. The events to which he refers took place in the Peloponnesian War, but his sentiments apply equally to Roman auguries. The question is, do or do not roosters crow to presage or announce a victory?

Whatever both nature and chance came to happen by its similarity sometimes gives rise to error; it is a great stupidity to consider the gods as the source of such events and not to seek their real cause. You believe the Boeotian soothsayers at Lebadia saw that by the crowing of roosters the victory had been won by the Thebans because conquered roosters are accustomed to keep silent but crow when they are victorious. Therefore, did Jupiter give a sign by means of chickens to so great a city? Or are those birds not accustomed to crow when they had not beaten some other rooster? This is, you will say, a portent. Indeed, a great one. As if fish, not roosters, had crowed.

It's Pliny who tells us that a woman can use chicken eggs to make sure that her pregnancy results in the birth of a son. The woman in his story is none other than Livia Drusilla (58 BC–AD 29), who was to become the third wife of Augustus Caesar. At the time, however, she was married to Tiberius Nero (not the Nero, born in AD 27, who fiddled while Rome burned). When she became pregnant, she was eager to bear a son and made her magic this way: She tucked an egg between her breasts; when she needed to put it down, she handed it to a maidservant, who would cherish the egg in the same way so that it would not lack warmth. Eventually, the egg hatched, and Livia's wish came true: She bore Tiberius Claudius Nero, who grew up to become Tiberius Caesar Augustus, Rome's second emperor.

Our most authoritative source of information on Roman chicken-keeping comes from Lucius Junius Moderatus Columella (AD 4–c. 70). After he left a career with the military, he took to farming and wrote about it at great length in a book named, like Varro's, *De re rustica—On Rural Subjects*. To our great good fortune, almost every last word of it has survived to give us a full picture of Roman agriculture from soils, vineyards, and olive trees to beekeeping and supervising employees. Book VIII deals with fish and fowl—ponds, geese, ducks, guinea fowl, and, of course, chickens.

Columella notes gamecocks and their training, but his mission is to promote raising chickens as a source of income for the farmer. They provide food for the table and dung to fertilize grapevines and trees and to enrich the soil. The farmer simply needs to determine how many birds he'll keep, and of what kind. Two hundred birds, Columella declares, are the upper limit for one poultry-keeper. Forego fighting breeds in favor of barnyard birds. Avoid white hens, he says, for they are delicate and don't lay well. He describes the ideal hen:

> Let your brood hens be of a red color, square-built, big breasted, with large heads, straight, red crests; white ears; they should be the largest obtainable which present this appearance and should not have an even number of claws. Those are reckoned the best bred which have five toes but without any cross-spurs projecting from their legs.

Five toes—three to the fore, and two aft—he is likely describing the Dorking type brought to Britain a good century before Columella took quill in hand.

His ideal cock is like that of Varro: a muscular bird with bright red wattles, golden hackles, brawny wings, well-feathered thighs, and not just *salax* but *salacissimus*—most salacious.

> These male birds, though they are not being trained for fighting and the glory of winning prizes, are, nevertheless, esteemed as well-bred if they are proud, lively, watchful and ready to crow frequently, and not easily to be frightened; for on occasion they have to act on the defensive and protect their flock of wives, nay, even to slay a snake which rears its threatening head.

Columella does not recommend bantams, unless you fancy smallness. Bantam roosters are extremely undesirable because, being super-feisty, they will attack other roosters and so keep them from treading their hens. To prevent these attacks, should you have a bantam cock, Columella advises putting a shackle made of leather round his leg.

Columella's kind of rooster »

For each cock, Columella prescribes five hens. For two hundred birds, that means some thirty roosters, but combat and crowing are reduced by housing them in one building and their hens in another. The inherited supposition that thunder spoils eggs pops up once again in Columella, but his remedy is more elaborate than that of Pliny: Put grass in the nest boxes, along with branches of bay, and use iron nails to fasten knobs of garlic beneath the boxes. He's very specific about coop design, giving precise measurements. Coops should have windows for good lighting, wicker nest baskets, plank ladders, and perches to keep the chickens from soiling themselves with their own excrement. His coops are constructed to keep out cats and snakes. The breath of a snake is said to kill chicks. He discusses laying and brooding, the food to be given to the flock, the poultry yard, diseases and their treatment, and keeping records of the eggs collected. Cleanliness is paramount to successful chicken-keeping.

Columella might have been appalled at the rise of urban chicken-keeping these days. He states at the beginning of Book VIII that raising fish and fowl is an activity suited only to the countryside and its farms. Farmers raise chickens for their own benefit, not that of people who live in town. But I'm willing to bet that he'd have found poultry in the poorer districts of Rome and maybe in the fancier ones, as well.

The survival to this day of all his books on agriculture is clear testimony that they were indeed useful and that their advice stood many farmers in good stead. We could still build a fine, well-lit, predator-free chicken coop according to his specifications. Our hens could nest and brood in wicker baskets. And we certainly still use chicken dung to enrich our gardens.

CHAPTER FOUR

The Medieval Chicken

The cock setteþ nexte to hym one rooste þe henne þat is
moste fatte and tendre.

John de Trevisa, 1398

A scene of domestic bliss—this is the picture given here of the cock perched on his roost next to the fattest, most tender hen, surely the favorite of his harem. Trevisa's English—þ equals *th*—translates the thirteenth-century Latin of *On the Properties of Things*, an encyclopedic work by Bartholomeus Anglicus, a Franciscan monk fascinated by all things animal, mineral, and spiritual.

The English words that we use for *gallus*—"chicken," "cock," "cockerel," "rooster," "hen," "fowl"—settled on the bird in the Middle Ages. "Chicken" was certainly spelled in a curiosity of ways. Old English had it as *cicen* for one bird, *cicenu* for more than one. Middle Low German and Low German spoke of the *küken* (you can spot the word *cock* here), while the bird was *kieken* or *kuiken* in the Netherlands. The *-en* is a diminutive. These words, which have no truck with the soft Romance languages, come from the Teutonic treasury of short, no-nonsense words.

Chickens were low-key, though essential, birds throughout the Middle Ages. They formed a large part of the diet for people of every class throughout Europe. So much is attested to by the prevalence of recipes for capons and pullets—roasted, boiled, made into soup—in two surviving recipe books from the late fourteenth century: *Le Viandier* by Taillevent, pseudonym of Guillaume Tirel, master cook for the King of France, and *Forme of Cury*, *Forms of Cooking*, by the chief Master Cooks of King Richard II. (For a recipe from the former, see the chapter on Chicken Cuisine.) But because of the birds' low-key profile, it's not easy to recon-

Dorking chickens, an ancient Roman breed »

struct the medieval chicken. We know that goodly numbers of them had been kept in Britain during Roman and Anglo-Saxon times. In fact, the Dorking chicken, a breed long associated with Britain, was probably introduced by the Romans. A broad-breasted, five-toed bird described by Columella, it comes in an array of colors from white and silver-gray to dark and cuckoo. The last indicates black feathers banded with white.

As a document from the tenth century attests, another breed was present in Scotland and northern Yorkshire in the Middle Ages—the Scots Dumpy. The word "dumpy" means short and stout, and short these birds are, for the legs of an adult are no more than an inch and a half long. As a result, Dumpies cannot run in the usual chicken fashion; they waddle, wherefore they are also known as Creepies or Crawlers. The birds were traditionally kept by crofters—farmers with small holdings, that is—in the Highlands or islands of Scotland. These crofts were set in wild places that offered perfect cover for predators. But Dumpies, swaying from one little leg to the other, were prevented from free-ranging at any distance from the croft. So, they stayed close to home under their owner's watchful eye. The bargain here was, I'll care for you so that you provide me with eggs, meat, and a clean-up service for

spilled food that might otherwise attract mice and rats. Legend has it that the Picts, an Iron Age and early medieval people living in northern England and Scotland, kept Dumpies in their war camps because the birds would give alarm when strangers approached.

« Scots Dumpy rooster:
Note the short legs.

Luckily a modern-day scholar has put together a picture of chicken husbandry in medieval England. The best-kept records for chickens there come from the great abbeys and priories, where keeping good accounts would be regarded as incumbent upon the servants of God. The records from English demesnes—manorial estates, that is—are, in comparison, hit or miss. Though the peasantry in Britain did not keep records at all, they certainly did keep chickens. The proof of that is that tenants on the estates paid chickens and eggs as rent for the land that they farmed. For both the monastic and the secular efforts, raising chickens was a labor-intensive enterprise in which expenses could easily outrun income. Nonetheless, an appetite for chickens and their eggs persisted, as it has done since chickens were domesticated and as it shall continue to do.

No factory farming existed in the Middle Ages. All poultry, including geese and ducks, was free-range. Of these birds, chickens were the most affordable, for they bore the lowest price when they were sold. They were economical in another way, for one cock could service many hens, while ducks and geese are monogamous. The people who tended the chickens and other fowl were dairy maids and sometimes children. They kept an eye out for predators, walked and supervised the birds, fed them with barley (shades of the Romans), made sure that the chickens were securely enclosed for the night, and drove them to market. In the egg department, the medieval hen was not nearly as productive as a modern hen. Pullets and cockerels were often sent to the table and eaten instead of being allowed to mature.

In England, a great shift turned chicken-keeping practices topsy-turvy in the middle of the fourteenth century. The Black Death, striking in 1348, killed some 40 percent of the population. Church and state suffered mightily. Prices fell, and real wages rose. The manorial estates did away with labor-intensive crops and animals, like chickens. Many farmed out their lands to the peasants, who, given the shortage of manpower, enjoyed an improving standard of living. The chicken rents were decreased or abolished, in part because fewer peasants meant fewer chickens. The recovery from the ravages of the Black Death brought a concomitant shift in what people ate, with the well-to-do favoring beef. But chickens hardly disappeared. As one chicken scholar says, "The vanishing of chickens from the seigniorial diet and the demesne does not necessarily contradict the archaeological evidence, which points that chickens were omnipresent both before and after the Black Death." The

« The cockatrice hatched by
human imagination

peasantry kept raising as many chickens as ever
it had.

In the twelfth century, the medieval imagination conjured up a terrifying chicken-headed
monster, the cockatrice. This creature hatched
out of an egg laid by a cock and incubated by—
the legends vary—a toad or a snake. In many
representations, its head and legs are those of a
fighting cock; its spiny, snakelike tail is tipped
with a triangular, fleshy tip that has a spine, like
a stingray's barb. With its leathery wings, shot
through with long bones, it could fly. One glance from its red eyes could petrify people. Or
if it so chose, it could kill them by touching or breathing upon them. The weasel was its
mortal enemy. Some believed that a cock's crow would cause it to die. The Medusa gambit
was the only sure way to best the beast: Let it look at itself in a mirror. But beware: Even after
death the head could turn you to stone.

The Bible, according to the fourteenth-century translation of the Bible by John Wycliffe's
team of scholars, gives room to the cockatrice in several verses. Middle English for Isaiah
14:29 puts it this way: "Al thou, Filistea, be not blad, for the yerde of thi smytere is maad
lesse; for whi a cocatrice schal go out of the roote of an eddre, and his seed schal soupe up a
brid," The post-medieval King James version of 1611 clarifies the verse: "Rejoice not thou,
whole Palestina, because the rod of him that smote thee is broken, for out of the serpent's
root shall come forth a cockatrice, and his fruit shall be a fiery flying serpent."

CHAPTER FIVE

The Renaissance Chicken

The Renaissance had its quota of chicken students and fanciers. Right or wrong, they were not in the least reluctant to repeat the opinions of their classical predecessors. One of the most delightfully credulous was Giambattista della Porta (1535–1615), a Neapolitan. A drawing shows him as Roman-nosed and balding but well mustached. In his narrow face, his eyes are dark and as baggy as

if he'd frequently stayed up late, squinting at his manuscripts by candlelight. He wrote a book on natural history—in Latin, of course, for that was the language of the learned. It was *Magiae naturalis*, which first appeared in 1584 and was subsequently translated into English as *Natural Magick* in 1658. It treats all manner of topics both natural and unnatural, including instructions on preparing compounded medicines, making seawater potable, kindling fire with a magnifying glass, making invisible ink, getting prey animals drunk so that they

Giambattista della Porta »

can be caught by hand, tempering steel, and beautifying women by dying their hair, applying makeup, and firming up saggy breasts. He also has much to say about husbandry and agriculture. But at book's beginning, he defines the meaning of "natural magic":

> There are two kinds of Magick; the one is infamous, and unhappy, because it has to do with foul spirits, and consists of incantations and wicked curiosity; and this is called Sorcery; an art which all learned and good men detest; neither is it able to yield any truth of reason or nature, but stands merely upon fancies and imagination. . . The other Magick is natural, which all excellent wise men do admit and embrace, and worship with great applause; neither is there any thing more highly esteemed, or better thought of, by men of learning.

So, della Porta deals only with that which he considers natural. He's given to such statements as "Mice are generated of putrefaction," and "Red Toads are generated of dirt, and of women's flowers."

But you want to know what he says about chickens. To begin with, he repeats Aristotle's erroneous dictum that "if the Egg be exactly round, then it will yield a Cock. But if it be somewhat long, then it yields a hen." One chapter in the fourth book bears the title "To Hatch Eggs without a Hen." He tells Aristotle's tale about the drunkard who sat drinking until the eggs under his mat hatched, and he describes the Egyptian hatching ovens, saying, again erroneously, that the eggs are covered with hot dung. And he tells how he tried to hatch eggs without a sitting hen by putting her dung into "a hollow vessel with a great belly." He turned them once a day for twenty days, as a hen would—except that a hen turns them more often. Alas! He writes, "I tried this most diligently, and it took no effect." He then gives elaborate instructions for building and using a hatching oven that will hold three hundred eggs. Nowhere, however, does he so much as intimate that he himself has succeeded in producing chicks by this method. Rather, he leaves the proof in the hands of others. But, if the magic of caring for new-hatched chicks can't be done any other way, then:

A Cock or Capon will perform what the Hen should. Do but show him the Chicken, and stroke him gently on the back, and give him meat out of your hands after, that he may become tame. Then pull the Feathers out of his breast, and rub him with Nettles. For in a few hours, not to say days, he will take care of the Chickens so well and give them their meat, that no Hens did ever do it as he will.

Judging by the number of culinary hints in the fourteenth book, della Porta relished eating not just chickens but also any meat fit for human consumption, from pig's liver and ox flesh to shell creatures. He gives several methods of making chickens tender. Hens should be dropped off a high tower. Because they shake their wings as they fall, they aren't hurt, but they are afraid, and fear serves as a fine tenderizer. As for a tough old rooster, "drowned him alive in Muscadel outright, and he will soon come to be tender meat." And, citing classical writers like Plutarch, della Porta offers yet another method for making roosters tender:

Wild cocks bound to a Fig tree, will grow tame and stand immovable. . . It is certain, as we may judge by sight, that the Fig tree sends forth a vehement and strong vapor. . . Wherefore the Fig tree sends forth a hot and sharp vapor, and that is digesting, and dries and concocts the flesh of birds, so that they grow tender.

Della Porta gives instructions, too, on fattening hens, a process that he calls "cramming." The birds, between five and twenty months old, are to be shut in their pens, each one with her head sticking out of one hole and her tail out of another so that movement is impossible. So, "they may eat their meat, and Shit it out again when it is digested." To prevent lice and soiling themselves with their droppings, all the feathers on their heads, thighs, and under-wings should be plucked. Soft hay should cushion their quarters, for without cushioning, they'll fail to grow fat. Then they are fed to a fare-thee-well. Nor does his culinary advice in regard to chickens end there. Here's how to roast a chicken without a fire:

> Put a piece of Steel into the fire, put this into a Chicken that is pulled and his Guts taken forth, and cover him well with cloths, that the heat breathe not out, and if he do smell ill, yet the meat is good.

As a man possessed not only of an adventurous spirit but also a raging curiosity, della Porta has few rivals. If we find his pronouncements about how to deal with barnyard fowl sometimes preposterous, we can certainly thank him for looking into many matters having to do with chickens. And we can note that he writes nothing about cocks as entertainment. He has listened instead to the dictates of his belly.

"It is clear to all, how much benefit the rooster and his wives provide for the human race." So said Ulisse Aldrovandi (1522–1605), the magisterial chicken man of the Renaissance. A drawing shows a stocky man wearing a robe with a straight yoke made of fur—he was a professor of philosophy and natural sciences at the University of Bologna. His balding pate is fringed with white hair, but his beard is full and well-trimmed, coming to a point below his chin. Like the fur from his shoulders, his luxuriant mustache descends on either side of his mouth. His most arresting feature is his eyes, wide open, filled with obvious curiosity, aware of everything around him. He collected botanical and zoological specimens by the thousands for a curiosity cabinet that he called his "theater." Some of the specimens still exist. He also imagined and directed the creation of a public botanical garden in Bologna.

⩘ Ulisse Aldrovandi

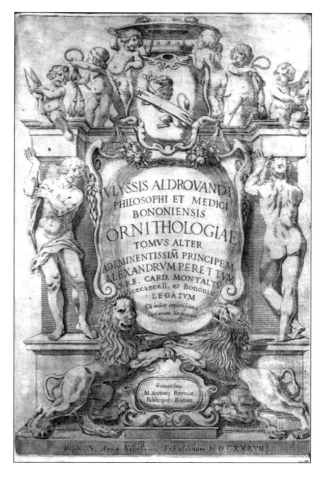

And he wrote books—compulsively, it seems—about all manner of creatures and plants. Written in Latin, they include learned tomes about the pharmaceuticals and other items in his herbarium, snakes and dragons, insects, and birds. He was a truly learned man, fluent in Latin, Greek, and Hebrew, as well as in several European languages. And he was able to cite his predecessors, from Aristotle through Columella to contemporary commentators. His books are lavishly illustrated with watercolors, many of them covering a whole page each, with descriptions tucked into the corners. To make these paintings, Aldrovandi paid artists out of his own pockets. Some of them stole their work and sold it.

It's Book XIV of his *Ornithologiae—Ornithology*—that cackles and clucks and crows. The name of its first chapter identifies the subject: "Concerning Domestic Fowl Who Bathe in the Dust—The Chicken Male and Female." The book's title page is fanciful, showing six *putti* who cavort at the top, and two muscular men who hold up the *puttis'* tablature and act as columns on either side of the names of the book and its author; at page-bottom, two lions with lush manes engage in rowdy play. But this page is not nearly so fanciful as some of the contents of the book. Despite a stated aversion to handing on unverified information, Aldrovandi occasionally fell victim to just that. At times, he was simply unable to resist the marvelous—and the stranger, the better.

Many pages of the chicken book are devoted to such subjects as fattening hens, cockfighting, chicken diseases, and egg-laying by roosters. The last he does not credit, despite many assertions from his predecessors that it is indeed so. He does believe many ancient assertions—that the gender of a chick can be discerned from the shape of the egg, that the crow of a rooster frightens a lion, and that thunder addles eggs. He also sees chickens as portentous: Roosters announce the coming of rain by crowing loudly after dark and flapping their wings; hens presage a heavy rain if they come home with their chicks as rain is beginning to fall or if they refuse to leave the hen house in the morning.

Hens are the model of motherhood. His admiration for them is evinced in his keenly observed description of the means by which they protect their new-hatched chicks:

> They follow their chicks with such great love that, if they see or spy at a distance any harmful animal, such as a kite or a weasel, the hens first gather them under the shadow of their wings, and with this covering they put up such a very fierce defense— striking fear into their opponents in the midst of frightful clamor, using

⊼ Mother hen with her new chicks

⩒ Aldrovandi's drawing of the oviduct

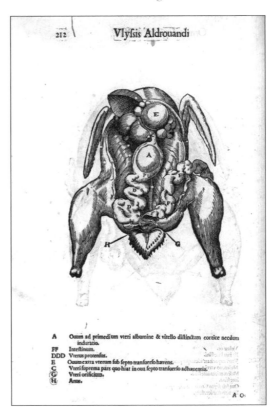

both wings and beak—that they would rather die for their chicks than seek safety in flight, leaving them to the enemy.

True science is evinced in Book XIV's illustrations of a hen's urogenital system. Led by a fascination with the bird's egg-laying apparatus, Aldrovandi dissected hens so that the single ovary, oviduct, and uterus were revealed. The drawings made from the sundered hens are as informative today as they were then. But then, knowing nothing of dinosaurs, he mentions the famous question:

> I pass over now that trite and thus otiose rather than curious question, whether the hen exists before the egg or vice versa. It is stated in the sacred books that the hen existed first. These books teach that animals were created at the beginning of the world; hence the hen did not come from the egg but from nothing.

We would recognize most of Aldrovandi's chickens as chickens indeed—the fat hens, the wattled roosters with great combs and flowing tail feathers. He describes various breeds in great detail: a petite bantam hen with deep red wattles and comb, the crested Paduan chicken with a pert, upright tail, a stub-tailed hen found in India, and multi-colored Turkish fowl. The Paduan chicken with its crest of feather bobbles is none other than today's Polish. The Turkish rooster shows his jungle fowl ancestry. Some of Aldrovandi's chickens are as black as possible or barred black and white, like our Australorp and Dominique breeds, respectively. One has blue

The green- and rose-colored Indian rooster »

legs and feet, like the modern Poulet de Bresse. Another has feathered legs similar to those of the Booted Bantam or Bearded d'Uccle.

In the realm of the stranger, the better, he presents a basically green- and rose-colored Indian rooster with a double tail, one of five short red feathers, the other of nine plumy feathers bearing eyes like those on a peacock's tail. Its upper bill is hooked like that of an eagle, and—Aldrovandi swears to this—it has ears. He introduces us, as well, to *Gallina lanigera*, another bird that he surely never laid eyes on: a "wool-bearing hen" covered with extraordinary curls that he writes are "not feathers but wool." He's also read Marco Polo's account of black hens, found in the city of Quelinfu in the kingdom of Magni, that have hair, not feathers. He reports that these Chinese hens "lay the best of eggs." It's likely that, in both cases, he is referring to the Silkie, a soft-feathered, fluffy breed.

Nor does he forget *Monstri*—freaks: a chick with three legs, a cock with four, two chickens with a single head, and a surely mythical rooster with *cauda quadrupedis*, the "tail of a four-legged creature"; it looks like a lion's sleek nether appendage with a fine bushy brush on the end.

Aldrovandi kept his own flock of chickens. He differentiated between tame and wild birds and, of the former, mentioned one with great fondness:

> Among the tame chickens are some so gentle and mild that they cannot live without human companionship. I have seen them with my own eyes. A few years ago in my country home I raised a hen who, in addition to the fact that she wandered the whole day alone through the house without the company of other hens, would not go to sleep at night anywhere except near me among my books,

Rooster with another animal's tail »

and those the larger ones, although sometimes when she was driven away she wished to lie upon her back.

Aldrovandi's approach to chickenkind was encyclopedic if not always accurate. But, because of his boundless enthusiasm and his wide-ranging curiosity, he can be forgiven his errors and cherished for the true interest that he shows for his feathered subjects.

It was in the Renaissance, that stellar age of seafaring exploration, that Europeans found chickens in every port. Ferdinand Magellan, the Portuguese circumnavigator of the globe (1480–1521), found sacred gamecocks in Borneo and Sumatra. Eating them was strictly forbidden.

Less than a century later, as Aldrovandi was writing his *Ornithology*, bantam chickens were introduced on a large scale to the west. Dwarfed chickens, like the Adrian breed, had been known and denigrated in Europe since classical times. But with the chartering of the great trading companies—England's East India Company in 1600 and the Netherlands' Dutch East India Company in 1602—to deal with the goods and resources of countries in the Pacific Ocean's Asian orbit, bantam birds came to the forefront. Ships put in at the port of Banten, capital of a sultanate on the northwest coast of Java. There the crews encountered miniature fowl that they named after the port, anglicizing it to Bantam. The birds were hardly brought aboard as pets; rather, they were foodstuffs. Bantams were not introduced to European chicken-fanciers until the nineteenth century, but it's possible to believe, given the appeal and friendliness of the little birds, that some bantams did indeed survive sailors' appetites.

Bantam chickens »

CHAPTER SIX

The Medicinal Chicken

Almost from the beginning, one of the bargains that we have made with chickens is: I will care for you because you heal me.

Wherever there are chickens, there is chicken soup. We all know that chicken soup is the premier panacea. It can cure anything that ails us. And so it has been throughout the association of human beings and *G. gallus domesticus*. Jewish mothers know its value full well, as do those who specialize in Chinese medicine, which can feature a soup called Wu Chi Pai Feng Wan made with many herbs and a plucked, charcoal gray chicken—a Silkie—including its shanks and feet. This soup's virtues are that it stops blood loss, increases your life span, and enhances fertility.

Ulisse Aldrovandi has much to say about the healing powers of the chicken. Nearly fifty pages of the English version of his chicken book are devoted to the medicinal chicken, with recipes that heal both internal and external ailments, from sore throats and belly aches to bites by poisonous snakes.

Silkie: Note the charcoal skin and beak »

The numerous afflictions he mentions are with us still, if not the same cures. "The genus of chicken," he writes, "offers so great an advantage to men in its use in medicine that there is almost no illness, both internal and external, which does not draw its remedy from these birds." As a polymath, he is also able to summon a thousand years' worth of commentary from old-time physicians and pharmacologists. Some, like Hippocrates (c. 460–370 BC) and Pliny, are familiar. Others have hardly become household words—like Gaul's Marcellus Empiricus (flour-

⌃ A bowl of chicken broth

ished AD 500) and the Persian physicians Mesue (AD 777-857) and Avicenna (c. AD 980–1037). Almost all of Aldrovandi's sources, no matter their era, swear by chicken broth.

A wonder of ailments can be cured with doses of broth, from constipation and flatulence to malarial fevers and leprosy. Perhaps even better, panthers and lions will not attack someone who has drenched himself in rooster broth, especially if it has been prepared with garlic. One recipe, said to be much favored in France, cooks an old rooster, with the foot of a calf or sheep, until the bird's meat falls off the bone. Then the juice is strained and seasoned with sugar and ground cinnamon. Egg whites and shells are then added before the mixture is again strained. If you wish, you can color it green or red, and if you'd like a sour broth, add a slosh of vinegar. Old-rooster broth acts as a palliative for asthma and heart trouble. Capons are used to make a broth that acts as a laxative. Broth can be made as well from young roosters and hens, but it only "tempers the harmful humors but does not draw them forth." Nonetheless, some medical writers swear by the broth of a young rooster as a cure for dysentery. If you want to know what to do with a gamecock killed in a drawn-out fight, he can be used to relieve malarial fevers in their acute stages. To render the gamecock medically useful, cook him with barley, seedless raisins, hyssop, thyme, pennyroyal, and violets; then mix vinegar and honey into the broth. Let the patient take as much as he can swallow in one gulp. Aldrovandi gives another means of preparing broth from instructions provided by the physician Dioscorides (c. AD 40–90), who wrote a five-volume book on medicaments. Dioscorides was still consulted in Aldrovandi's day. Here's his ancient recipe for rooster broth as given by Aldrovandi:

> Throw away the entrails and add salt. Sew up the stomach and boil the bird in twenty pints of water until it is reduced to half a pint. Cool the concoction and give it daily to the patient. Some cook marine vegetable, dog's mercury, safflower, or polypody (rock fern) along with it.

"Marine vegetable" means seaweed, and dog's mercury is a foul-smelling, acrid-tasting plant found in woodlands. Aldrovandi quibbles with Dioscorides for not giving the precise amount of salt and not specifying whether the rooster was to be cooked whole or in cut-up pieces.

It's not just rooster broth that works miracles. Eggs and eggshells also have their pharmaceutical uses. Aldrovandi quotes Pliny to the effect that a nosebleed can be healed with eggshell ashes soaked in wine. And our Bolognese natural historian also gives instructions that he learned from the physician and surgeon Galen (AD second century) on compounding another remedy for nosebleeds. It consisted of eggshell and the juice of unripe grapes rubbed together in water. The mixture was then to be applied to a piece of twisted linen that had been wetted with vinegar. The linen was then to be placed on the nose or forehead of the patient and covered with mud. The physician would stop up the ears of the patient. Cure effected. Aldrovandi cites a coagulant made of eggshells that is cited by the Swiss physician and naturalist Conrad Gessner (1516–1565), who wrote *Historiae animalium—The History of Animals.* Its third volume treats birds, including a hundred pages on the chicken:

> Grind up eggshells in sharp vinegar until they are softened. Dry them in the sun and scatter them over the wound wherever the blood flows; or scatter dust mixed with miller's soot and it will soon stop.

We'd have trouble replicating this remedy today, for mills with sooty chimneys are few and far between. But never fear. The ashes of eggshells from which chicks have hatched serve to reduce kidney stones. Burnt eggshell was also a particularly potent remedy. Galen wrote, "Burn up the entire shell of an egg and mix the liquor extracted from it with cracked arsenic. Pour it into the nostrils of the patient. If arsenic is not obtainable, the liquor of the egg alone will suffice." (This concoction sounds as if it might kill before it managed to heal.)

Egg whites have a wonderment of uses. Pharmacists used it to clarify syrups. Other recipes for egg whites were not medicinal. Aldrovandi quotes Pliny's mention of mixing egg whites with quicklime to glue together pieces of glass. This glue's "strength is so great that wood soaked in egg white does not burn nor does clothing which has been treated with it." Albumen was also an ingredient in the cement used to bond together bits of stone in mosaic work. It figured, too, as a prime element in making paint. Aldrovandi writes, "Those who decorate a picture with color break the egg white with a sponge until it becomes thin and watery. Then they mix it, thus broken, with their colors, as our Italian painters do." He adds, almost as an afterthought, "Once egg white was used for adorning and curling the hair by young men; now it is so used only by the girls."

If you're hoarse because you've shouted too loudly and long, or if you have a sore throat, try sucking down an egg white. Newly laid eggs are said to work best because they still retain warmth. To reduce a fever, Hippocrates, the eminent Greek physician, recommends drinking three or four egg whites beaten in an amphora—about six pints—of water. Aldrovandi comments that Hippocrates "promises that this mixture will greatly chill the patient and disturb him to the point where he empties his bowels." Lukewarm eggs are recommended to cure shortness of breath and pleurisy. Pliny offers an inebriating cure for dysentery: five raw egg yolks in half a pint of wine, along with broth made from the eggshells, poppy juice, and more wine. (Yes, poppy juice is opium.) When it comes to tuberculosis, its grating cough and discharge of bloody mucus can be eliminated by giving the patient a mixture of raw egg, the juice of a cut-up leek, and an equal amount of honey. Cooked in vinegar, eggs heal ulcerated kidneys and bladders. An egg yolk by itself, eaten raw, helps with the same problems.

Marcellus Empiricus provides an egg remedy to reduce fevers and expel intestinal worms:

> Take a raw egg open at the top, fill it with green oil and pour it out; then fill it with the urine of a virgin boy and pour this out. Then add a little honey, mix with the inner parts of the egg itself, and give the result of all these operations to the fasting patient. It will drive out the oldest feces and noxious worms and relieve the most acute fever.

No part of a bird goes unused in the chicken pharmacy. A whole bird, rooster or hen, with its rear body feathers and fluff plucked, can be placed with its anus over a wound to suck out the poison. Rooster brains are of great help in treating cerebral palsy, shock caused by the bite of a poisonous creature, and, indeed, lunacy. Hens' brains will stop a nosebleed. A rooster's dried windpipe drunk in warm water by a fasting patient cures incontinence. A mere spoonful of this medicament prevents bedwetting. Rooster testicles, either eaten dry or, as Pliny recommends, swallowed with water, preceded by milk, will serve not only as a remedy for epilepsy but as a stimulus to a lusty erection. Pliny also offers the opinion that if a woman eats testicles right after conception, she will surely bear a male child. The still palpitating heart of a newly slaughtered rooster strapped to a woman's pregnant belly will speed up childbirth, though Pliny offers another prescription saying that birth will be accelerated if an egg seasoned with rue, cumin, and anise is drunk in wine. (Alcohol as anesthesia?) A rooster's gizzard, dried, crushed, and administered with a goblet of wine, relieves a cough. His liver, mixed with poppy

juice, cures colic. A feather dipped in vinegar and brushed across a patient's nostrils will immediately bring him out of a faint; in sharp contradiction, it is also said to induce deep sleep, as does eating rooster feet. The neck feathers and crest of a rooster tied around your head will banish a headache. Here, according to Marcellus Empiricus, is what you can do with chicken fat to heal abcesses:

> Three spoonfuls of the powder of old dry anise and the powder of pine-cone resin with the fat of an old chicken must be eaten in the morning without any other food by the patient suffering from abcesses; give him the same amount at evening, and you will help him wonderfully.

Chicken legs, complete with their feet, cooked and eaten with salt, oil, and vinegar, will dispatch colic.

Dung needs a short paragraph of its own. Drinking dung mixed with vinegar or wine relieves colic. Conrad Gessner refines this prescription with a recipe he read in a German manuscript: Use only the white part of the dung. For another pharmaceutical, Aldrovandi cites the work of Arnaldus of Villanova (c. 1235–c. 1312): If you need a purge, two drams of young rooster dung dissolved in warm water makes you vomit. (Two drams equal an eighth of an ounce.) Aldrovandi goes on to say that some people think that ingesting the white part of a hen's dung will cause coagulated blood to be expelled by vomiting. (The very idea makes me feel queasy.) From a thirteenth-century Byzantine work on medical materials, Aldrovandi notes white rooster dung as a treatment for angina. He does not mean the dung of a white rooster but rather the white part of the dung. His source advises laying aside dung:

⌄ Producers of dung

[that] has dried and become the color of white lead. The dung is to be administered with water or mixed with water mead and offered in a spoon; he promises that
this treatment will cure desperate cases. If the patient cannot drink it, he advises that
it be mixed with honey and smeared on the inner parts of the body.

So far, except for using a whole chicken to draw out the poison of a snakebite, these remedies
all address internal ills. Chickens are also used to cure a host of ills—burns, itching, boils,
carbuncles, pimples, ulcers, erysipelas (a skin infection caused by streptococcal bacteria),
pinkeye, arthritis, earaches and toothaches, gout, breast tumors, wounds, snakebites, and
spider bites. The most important items in the pharmacopoeia are—you guessed it—eggs and
dung. Raw eggs remove the pain of burns and may keep blisters from forming. They make a
good medication for hemorrhoids, as well. Egg whites serve to coagulate wounds. Direct
applications of dung are especially efficacious for bursting boils, though the experts argue
about whether it should come from a red rooster or a white one. Poor vision may be made
better by smearing the dung of a red rooster on the eyes, while others use white-rooster dung
to stop tears. Mix white chicken dung with vinegar and honey, then drink it to reverse the
effects of mushroom poisoning. If you've been bitten by a rabid dog, madness can be kept at
bay by eating chicken dung with your food. Apply dung reduced to ashes on leg ulcers, and
they shall heal. Put dung ashes on your feet to heal ulcerations there. And if you have male
pattern baldness, dung smeared on the scalp can prevent further loss of hair.

(A note on contemporary dung: Simone Chickenbone, LLC, located in Wichita, Kansas,
manufactures a lip moisturizer called Free Range Chicken Poop Lip Junk. Its ingredients
comprise avocado oil, beeswax, jojoba oil, lavender essential oil, sweet orange essential oil,
and vitamin E. A disclaimer on the package reads, "Contains No Poop." Simone Chickenbone may very well be missing out on a good thing.)

Chickens also furnish other well-regarded simples. The blood restores vision to the blind;
it heals erysipelas and chilblains. Liquid chicken fat put in the ears will vanquish an earache.
To soothe the pain of teething, coat a baby's gums with solid chicken fat. Drill a hole in the
last bone in a chicken's wing, bind it with a seven-knotted thread, and suspend it over a
swollen arm or leg—the swelling will subside. Aldrovandi presents Pliny's prescription for

banishing a toothache: Use preserved hens' bones with the marrow intact to touch the tooth or scratch the gum. Pliny gives a recipe for dental health, as well: Put ash of eggshell in wine and use it as toothpaste.

In Aldrovandi's chicken book, the section on the birds' medicinal uses concludes with several pages devoted to veterinary medicine. Horses as well as people can be helped by the chicken pharmacopoeia. Once again, dung plays a prominent role. It alleviates the pip, heals wounds, and rouses the libido of a mare that has theretofore refused to be mounted by a stallion.

The desperate human struggle against disease is all too evident here. It led people to embark on what, today, seem like extreme measures. No need for us to raise cure-all chickens. We have drugstore ways of dealing with such ailments as constipation, diarrhea, acne, pimples, and aching heads, ears, and teeth. We've made headway, too, with more serious diseases, like arthritis, tumors, abscesses, angina, and much else.

But we still need chickens. They are used today in new technologies. See the chapter Chicken Science.

CHAPTER SEVEN

The Transitional Chicken

Enter unnatural history, though of a benign sort. Swift human intervention takes the place of nature's slow windings. And, oh, the breeds that we have engineered! We have tampered with chicken genetics since prehistoric times. One study says that "chicken breeding represents one of the most remarkable examples of directed evolution." The Department of Animal Science at Oklahoma State University lists sixty-three breeds of barnyard and pet chickens in the United States and Europe, which fall

into three modern categories: broilers for meat, layers for eggs, and the rest as ornamental, good for nothing but being strange and/or beautiful. The Department of Animal Science does not include varieties that exist only in China, Japan, and places elsewhere in the world. Breeds like the jungle fowl and the long-tail cock have stayed in their Asian homes. Others, like Cochins and the little Silkies, came to us from the Far East.

Naked Neck chickens »

Some of our recognized breeds are familiar, like Rhode Island Red, Leghorn, Buff Orpington, and Plymouth Rock. Others take eccentricity to various extremes. One is the Appenzell Bearded Hen, which sports beak-surrounding feathers that look for all the world like mutton-chop whiskers. Another, the Naked Neck or Turken, wears only a small tuft of feathers near the collarbone on its otherwise denuded neck. The Crèvecoeur was bred not as food but as an ornament; the heads of these birds are crested with fine black feathers that look like a helmet, and their bodies are heavily clad in black with a shimmering green iridescence. The Silkie breed is also an ornamental—like any chicken, it is edible, but its charcoal-black skin puts off would-be diners in the Western world; they are relished as food and medicine in the Orient. Their feathers, lacking barbules—the hooklets that lock feathers together—distinguish them; grown birds are as soft and fluffy as chicks. Faverolles, both roosters and hens, wear beards and sprout feathers on the backs of their legs. Some chickens, like the Modern Game breed, are skinny, long-legged creatures that look as if they walk on stilts, while others, like the Jersey Giant, are extravagantly plumped out. We've had as fine a time messing with *G. gallus* as we've had creating weird and wonderful variations of dogs and cats.

More than that, chicken breeds have been assigned to classes according to where they were developed. Members of the Asiatic Class include Cochins and Chinese Langshans with leg feathers, fighting Aseels, and Sumatras with double spurs. The last were imported into Britain in 1847 with the notion that they would be superior gamecocks. That, to the chagrin of sportsmen, is the year that cockfighting became illegal in England, which is not to say that it disappeared. The Oriental Class includes the Silkie and the Shamo, a long-legged game bird developed in Japan from Thai ancestors.

The Japanese Shamo »

Andalusian rooster and his hens »

Closer to home, the Continental Class includes, among others, the Polish with floppy crests; the Dutch Barnevelder, which lays dark-brown eggs; and the Belgian Campine with both hens and roosters wearing a golden cape. Among the Mediterranean Class are Leghorns, developed in Tuscany, and the Spanish Andalusian, which Gregor Mendel used for studies in genetics and heredity. None of the Mediterranean birds is commonly found in the United States. But those of the English Class are: the Orpingtons that come in many colors, the Dorkings that Britain has known since the time of Julius Caesar, the big Black Australorp, and several others.

Much of this chicken engineering took place in the transitional period that Page Smith and Charles Daniel, authors of *The Chicken Book*, have called the century of the chicken. The century began in the late 1700s and lasted throughout the 1800s. It was a time of clinging to classical ideas and the dicta of Aldrovandi, though the modern chicken was already peeping in its shell. In the 1800s, chickens were not only birds widely kept for their eggs, meat, and company, but they also became show-worthy because of their plumage, their conformation, and their sheer, elegant quirkiness. A new bargain was made: I will care for you if you beat out the competition and win me prizes. The familiar breeds became standardized, and totally new variations on the chicken theme were introduced with the import of Oriental chickens, which looked nothing like the five-toed British Dorkings and the Dominiques, a breed that originated in America. The first American poultry show was held in Boston in 1849. The chicken left the coop and the barnyard and entered the realm of high style and gaudy feathers.

The year before the Boston show, an Englishman, the Reverend Mr. Edmund Saul Dixon, Rector of Intwood-with-Keswick, published an influential book that brought joy to the hearts of chicken people—*Ornamental and Domestic Poultry: Their History and Management*. Its preface says:

Dominique, an American breed »

The history of the present volume is very simple, and, it may be, runs parallel with that of many other works on higher subjects. The Author, with his Wife (now removed from worldly trouble) and his Child, were living in a small suburban house, that had a little back garden attached to it. As a harmless amusement they procured a few Fowls to keep, although totally ignorant of their ways and doings. In aid of this ignorance books were procured—to little purpose. The difficulty of obtaining instruction from others led to close observation and a more eager grasp at the required knowledge.

Natural historians like Aristotle, Pliny, Columella, Giambattista della Porta, and the immensely learned Aldrovandi are cited frequently in the Reverend Mr. Dixon's work in regard to such matters as coop orientation (let it face south) and fattening hens for the table. He shares his predecessors' admiration for chickenkind throughout. He speaks of the hen as "deservedly the acknowledged pattern of maternal love." And here is his encomium to the cock:

> The courage of the Cock is emblematic, his gallantry admirable, his sense of discipline and subordination most exemplary. See how a good Game Cock of two or three years experience will, in five minutes, restore order into an uproarious poultry-yard. He does not use harsh means of coercion, when mild will suit the purpose. A look, a gesture, a deep chuckling growl, gives the hint that turbulence is no longer to be permitted.

The Reverend Mr. Dixon also treats of the traditional chicken-based medicaments that were used for a good two thousand years before he took pen in hand. They include several recipes for broth. Here's one that may have been inherited from Aldrovandi:

> Cock-broth is thus made: Tire an old cock till he fall with weariness, then kill and pluck him, and gut him, and stuff him with proper physic, and boil him till all the flesh falls off, then strain it. This broth mollifies, and by means of the nitrous parts wherewith that decrepit animal is endued, and which are exalted by the tiring of him, cuts and cleanseth, and moves the belly.

This broth, concocted from a weary and decrepit bird, is said to be a fine remedy for colic, cough, and tartar of the lungs. Nor did the Reverend Mr. Dixon resist dung—the white part of hen's dung, to be precise. He discovered this reference in the *Ornithologia* by Francis Willughby (1635–1672). Willughby's book, written in Latin in accord with seventeenth-century scientific practice, states: "*The Dung* . . . cures the Colic and pain of the Womb. Moreover, it is good especially against the Jaundice, Stone, and Suppression of Urine." The Reverend Mr. Dixon also offers other equally ancient remedies, like this one: "*The Weasand* of a Cock, burnt and not consumed, given before supper, is an antidote to the influence of the herb Dandelion." This medicament can be minimally clarified by translating "weasand," which means "stomach." As for the herb dandelion, it's often used as a remedy for snakebite. The need for an antidote may arise with liver failure: A little dandelion juice benefits, but more than a little is toxic.

The Reverend Mr. Dixon describes twenty-three breeds of chickens in delightful detail. Some are exotic—Malay, Cochin, and Silkie. Others originated much closer to home—Dorking, Spanish, Spangled Hamburgh, Polish, and Dominique. After surveying the large breeds, he has much to say about bantams, which he calls "the most treasured pets of the Fancy."

⋟ Francis Willughby

Spangled Hamburgh chicken »

We have advanced with a tolerably steady footstep through the flocks of well-sized creatures that crowded beside our path—the Turkeys, the Peacocks, the Geese, and the Swans—and should not have feared to encounter even an Emeu or a Cereopsis [genus-name of the Cape Barren goose], had chance planted one in our way; but a sudden fear creeps over us as we draw near to these mysterious elves and pigmies of the feathered world. Gulliver got on very well in Brobdingnag, so long as he did not attempt any leap beyond his strength; but the minute Lilliputians teased him sadly by their numbers, their activity, and the unseen and unsuspected places from whence they issued.

Bantams as teasing Lilliputians—a delightful analogy, and one that seems accurate in describing how some people are hopelessly entrapped by the fancy.

A few breeds mentioned by the Reverend Mr. Dixon seem to have disappeared off the face of the earth—Blue Dun, Jersey-Blue, and Shakebag. That last—what a peculiar name! Here are his observations on the bird, which never represented a breed but was rather named for an occasion:

THE SHAKEBAG FOWL, Commonly called the Duke of Leed's breed, is said . . . to be extinct; if so, it will not be necessary to consume much of the reader's time in describing a Fowl which he may never see. As a mere matter of history, we may state that the Duke, being an enthusiastic Cock-fighter, was in the habit of bringing his Cocks into the Pit in a bag, against any that could be produced, and, when shaken out, from their superior strength and size, were found more than a match for any competition, and were subsequently denominated Shakebags.

The book is jam-packed with such accounts of marvels and wonders. One small section deals with the possibility of hatching two chicks from a double-yolked egg. The answer was—and still is—no. There's not enough room in the shell to allow the full development of two little birds. Another section treats of a game played with chickens. Take a cock by the legs and whirl him around three or four times; when he's set back on the ground, he will stagger and reel like a drunkard. Under the name "Tipsy Hen," the same game may be played with the female of the species. The book also assesses the future of raising chickens for profit. The Reverend Mr. Dixon foresees the advent of companies that aim to mass-produce chickens, but he shakes his head at such an impractical notion—it's a bubble that will burst. "Meanwhile, we, prophets of evil, premise that the more densely poultry are congregated, the less profitable will they be; the more thickly they are crowded, the less will they thrive." He espoused letting them run free.

In 1850, J. J. Kerr, a physician living in Philadelphia, edited the Reverend Mr. Dixon's 383-page book, renaming it *A Treatise on the History and Management of Ornamental and Domestic Poultry.* Under his editorship, the book expanded to a monumental 509 pages. He added copious illustrations of the birds that it mentions, like "the several kinds of Shanghaes, Guelderland, and other varieties." He also had a great deal to say, much of it speculation, about the origin of chickens. You can hear an audible sniff in this comment:

> Those authors who, by a pleasant legerdemain, so easily transform one of the wild Indian *Galli* into a Barndoor Fowl . . . write as if they had only to catch a wild-bird in the woods, turn it into a courtyard for a week or two, and make it straightway become as tame as a Spaniel.

Dr. Kerr also lamented the absence of domestic chickens in the Old Testament. He was certain, though, that they were kept in Israel. In his opinion, this absence reflects the probability that tending the birds was women's work, not nearly so worth noting as "the active enterprises of men." He wonders about the earliest date of chicken-keeping and acknowledges that no one really knows. But he has a pretty firm idea that it includes Japhet, a son of Noah.

My own belief is, that it is coeval with the keeping of sheep by Abel, and the tilling of the ground by Cain—a supposition which cannot be far from probability if there is any foundation for the legend that Gomer, the eldest son of Japhet, took a surname from the cock. Indeed, it would be to him that Western Europe stands indebted for a stock of Fowls from the Ark itself.

Gomer's descendants left Asia Minor and, according to legend, became the Gauls, the Germans, and the Celts. Though no eyewitness testimony exists, they must, in Dr. Kerr's view, have brought chickens with them. But in his pre-Kon-Tiki age, the good doctor wonders how chickens had made their way to islands all over the Pacific, though he has heard through missionary sources that men in big native canoes could indeed make the journey from Sumatra to Tahiti. Explorers like Captain Cook noted the presence of fowl, along with pigs and dogs, "on islands that had never before been visited by civilized Man." Since prehistoric times, the chicken has been a world-traveler.

Both the Reverend Mr. Dixon and Dr. Kerr rely heavily on the views of Aristotle, Varro, Pliny, and Columella. Along with sure facts, they faithfully transmit classical misinformation: Thunder addles eggs. You can tell the sex of an unhatched chick by the shape of the egg. Avoid white hens because they don't lay well. Chicken studies have not yet truly cut the apron strings tying them to the classical past.

Ah, but those were the days! You could feed a laying hen on the best corn for two cents a week.

George Burnham (1814–1902), who raised a few chickens at his home in Roxbury, Massachusetts, was quite taken aback by the astronomically escalating prices for fancy birds engendered by the 1849 Boston Poultry Show. His shock at what he saw happening in the

Captain James Cook »

chicken world prodded him to write *The History of the Hen Fever*, which was published in 1852. Its first words are, "Never in the history of modern 'bubbles,' probably did *any* mania exceed in ridiculousness or ludicrousness or in the number of victims surpass this inestimable humbug, the 'hen fever.'" Its chapters trace the expensive course of this disease from its first symptoms to a final Shanghae dinner, in which all of Burnham's now worthless birds were served up broiled, stewed, roasted, curried, and made into pie and pudding. "The Hen Trade," Burnham wrote, "is a *fowl* calling." And that marks the finale for the transitional chicken.

CHAPTER EIGHT

The Modern Chicken

The modern chicken really began to come of age with the publication in 1857 of *The Illustrated Book of Domestic Poultry*, edited by Martin Doyle. It does not hark back to classical and Renaissance times. Its primary concern is the maintenance of healthy purebred flocks for eggs and meat.

The 1860s were bonanza years for chicken studies. In 1868, Charles Darwin (1809–1882) published *The Variation of Plants and Animals Under Domestication*, in which he investigated the origins and development of dogs, cats, and goldfish as well as of a slew of agricultural animals, from horses and asses, sheep, pigs, and goats to rabbits, pigeons, and fowl. Particularly fond of pigeons, he devotes two chapters to the various breeds, among them the Pouter with its huge crop, the Carrier, the Fantail with tail feathers that spread out like a peacock's, and the Runt, a misnomer for a really big pigeon. In the plant department, he surveyed everything from cereal grains and vegetables to fruit trees and flowers. It

Charles Darwin »

Charles Darwin.

« **Polish fowl illustration in Darwin's** *Variation of Plants and Animals*

is, of course, Chapter VII that concerns us here: his inquiries into the origin of chickens.

Darwin begins by listing breeds—Malay, Cochin, Spanish, Polish, Hamburgh, Silk fowls (Silkies), others—and sub-breeds, like Sultan, Houdan, and Crévecoeur. He notes the Japanese provenance of Bantams and the presence of Frizzles in India. As for chickens that lack a pope's nose, more precisely known as a pygostyle to which the tail feathers are attached, he writes with almost audible outrage: "Rump-less fowls—they are so variable in character that they hardly deserve to be called a breed. Anyone who will examine the caudal vertebrae will see how monstrous the breed is."

Darwin then explores the pros and cons of attributing the extremely diverse breeds that he sees in his day to a single parent-species or to more than one ancestor. To follow his debate, with each side carefully outlined, is like watching a man caught in the middle of a tug of war. He is pulled this way and then that. Listen to part of his internal dialogue:

> We see that the several breeds differ considerably, and they would have been nearly as interesting for us as pigeons, if there had been equally good evidence that all had descended from one parent-species. Most fanciers believe that they are descended from several primitive stocks. The Rev. Mr. E. S. Dixon argues strongly on this side of the question.

Every last kind of pigeon springs from the wild rock dove. Darwin seeks an equivalent ancestor for the chicken. In the end he settles on *Gallus bankiva* (the subspecies of red jungle fowl found in Java) as the primitive stock from which game fowl are descended. *G. bankiva* and game fowl can interbreed, and they bear a remarkable resemblance to each other. But whence come "the black stately Spanish, the diminutive elegant Bantam, the heavy Cochin with its many peculiarities, and the Polish fowl with its great top-knot and protuberant skull"? Anatomical examinations of the skulls of different breeds do not solve the problem. I can imagine a frustrated Darwin lamenting—no, damning—the absence of a single progenitor.

In 1867, the year before Darwin's research into plant and animal variations, a stellar no-nonsense book appeared: *The Poultry Book* by William Bernhard Tegetmeier, a British naturalist and *Gallus* enthusiast. Well acquainted with Darwin, he subscribed to Darwinian principles. His book contains no recipes for broth, no mention of ash of eggshell, and no whiff of curative dung. It refers to the work of its predecessors, like Columella and Aldrovandi, only when it is necessary to cite them for purely historical reasons, such as naming the varieties of Roman fowl and discussing the Polish bird, which Aldrovandi called the crested Paduan. The book lists one kind of chicken that is not assigned to any breed: "The title of Barn-door Fowl is given to the mongrels that are found existing in all places where no care whatever is taken respecting the purity of the breed of poultry." The book concludes with an appendix, "The Standard of Excellence in Exhibition," which lists with the utmost precision exactly how, from beak and comb to toes, tail, and coloration, the hens and cocks in each of thirteen breeds should look. Standards for turkeys, guinea fowl, and ducks are given, as well.

Tegetmeier's work predates, though not by much, the founding in 1873 of the American Poultry Association (APA), the oldest livestock organization in the United States. In 1874, it published *The American Standard of Excellence*, which described in encyclopedic detail the forty-one breeds that it recognized. The book's name changed to *The American Standard of Perfection* in 1905. The APA's books on standards are filled with drawings of not only the birds but also the color patterns of feathers, types of combs, and the nomenclature associated with both roosters and hens. The associa-

≽ A Cochin chicken

tion's goal was—and is—to keep out the poultry equivalent of mongrels by assuring pure bloodlines for barnyard fowl and fancy show birds, including ducks, geese, and turkeys. The first edition listed forty-one breeds of chickens, including the Dorking that arrived in Britain with Caesar. The Scots Dumpy, now so rare as to be considered endangered, never made the cut. The 2010 edition, a hefty hardcover volume with color plates, recognizes more than sixty breeds. It sets guidelines for shows and mounts them. Its standards are extremely conservative, as is only proper for an organization dedicated to purity of form, color, and line. The defects that it lists are legion and serve to take points off the score of a show bird: frosted comb, coarse texture of comb, an injured or missing eye, deformed beak, broken wing feathers, split wings, bow legs and knock-knees, crooked toes, and shanks, feet, and toes of a color foreign to the breed. Such stringency is deemed necessary to prevent inheritable defects from showing up in offspring. And the APA lobbies for a fair shake for poultry hobbyists, from the exhibitors to the people raising backyard chickens. One issue has been the shipment of day-old chicks, which airlines and FedEx have refused to transport. So the U.S. Postal Service, for all its vagaries, is the carrier that delivers chicks to all destinations. Deeming shipment in a box with air holes to be cruel and unusual punishment, PETA is incensed. But the voices of the hatcheries ring louder.

The American Bantam Association (ABA), founded in 1915, also sets standards for its petite birds. It lists fifty-seven breeds and eighty-five patterns for feather coloring. Like the APA, it sponsors shows and sets exhibition standards. As for their charms, two British writers say:

> Bantams are great "time-wasters." Five minutes allocated to feeding and watering can very quickly slip into half an hour as you watch them dust and peck their way across the garden while chirruping in the manner of buxom ladies doing the rounds of market stalls.

In the chapter on fowls in *The Variation of Plants and Animals Under Domestication*, Charles Darwin defined bantams: "BANTAM BREED—originally from Japan; characterized by small size alone; carriage bold and erect. There are several sub-breeds, such as the Cochin, Game, and Sebright Bantam, some of which have been recently formed by various crosses."

Darwin based his opinion that "dwarf fowl" had originated in Japan because they were referred to in an old Japanese encyclopedia. Most of the large breeds come in mini-versions— the feather-legged Cochins and Brahmas, clean-legged Araucanas, Rhode Island Reds, Wyandottes, and most of the others. The tiny, bouffant Silkie has always been a Tom Thumb among chickens. Other bantams are considered true bantams, for they have no large counterparts. They include the Belgian Bearded d'Uccle (pronounced *dew-clay*) and the Booted Bantam, both with elaborately feathered legs; and the Japanese bantam known as Chabo, which has very short legs that are sometimes invisible because they are covered by the thigh feathers. Other true bantams are the Rosecomb, with a backward-pointing comb that looks like a small red rapier; and the Nankin, said to have been in England since the 1500s. If so, it must have traveled a silk road from the Chinese city of Nanjing.

The Sebright, a special true bantam, is worth a few words. It was named after its developer Sir John Saunders Sebright (1767–1846), who hybridized these bantams in 1810 and started the Sebright Bantam Club for fanciers shortly thereafter. It was the first individual breed club for any kind of chicken. No one now knows exactly what breeds he used—Hamburgh, Polish, Nankin?—before he achieved birds that would propagate truly. Golden-laced and silver-laced, what beauties they are! The lacing refers to the slim black penciling on the margin of each feather. The Reverend Mr. Dixon turned up his nose at them, believing that Sir John had imported them from India, and called them impudent, damning the cock as a "little whipper-snapper! Pretty, certainly, and smart, but shamefully forward in his ways." But Darwin, an admirer, writes: "The male and female of gold- and silver-laced Sebright Bantams can be barely distinguished from each other, except by their combs, wattles, and spurs, for they are colored alike." In this case, the hens are as highly colored and gorgeous as the cocks. Sebrights were accepted as a breed by the APA in the very first edition of its book *The American Standard of Excellence.*

Chickens are not the only domestic birds that come in mini sizes. You'll also find bantam ducks, of which the

A golden-laced Sebright bantam »

⌃ Call duck, a bantam breed

Reverend Mr. Dixon has this to say: "A much smaller race of White Ducks is imported from Holland; their chief merit, indicated by the title of Call Duck, consists in their incessant loquacity." The color of a Call duck is not confined to white; their feathering may be blue with a white bib; buff; pastel, the male of which resembles a mallard; snowy, with cinnamon speckles; and gray. All have orange legs so bright that they're almost fluorescent.

The APA and ABA cater to people who keep fancy fowl and show them. They do not set the standards for the chickens and eggs that we buy at the grocery store. In fact the poultry associations would likely use Tegetmeier's term and call them "mongrels." We arrived at this mongrelization because of the Western world's Agricultural and Industrial Revolutions. The former increased food production on a fairly large scale; the latter severed many people from their rural lives and brought them in great numbers to work in factories. Their wages bought sustenance from the improved farms. At the same time, the popular appetite for chickens burgeoned, in large part because the birds, once luxuries, and their eggs had become affordable. In the last years of the 1800s, livestock, including chickens, became the objects of serious scientific study as universities inaugurated departments devoted to farm animals. By 1910, fully sixty-five universities and experimental stations were conducting research on poultry. Within the next few years, dozens of journals catering to chicken-keepers hit the newsstands.

The industrialization of the chicken got off to its small American start in the 1880s. The place was Petaluma, California, then a provider of fresh vegetables, meat, dairy products, and eggs to San Francisco. The person was Christopher Nisson, a Dane who had come to work there as a nurseryman. He noticed with great disapproval that many of Petaluma's chickens just hung out at barns and laid eggs under porches, eggs that just sat there until some farmer's wife eventually collected and sold them. Nisson thought that the time his hens spent on brooding and hatching their chicks was time wasted. The eggs should be available for sale in the big city. At that time, a small commercial incubator capable of hatching ten eggs had been put into production. Though some people scoffed, Nisson bought the newfangled thing and made it work. (No one then remembered the ten-thousand-egg hatching ovens of the Egyptians and Chinese.) He later built an incubator that had a capacity much greater than a mere ten eggs. Taking a tip from his successful enterprise, several Petalumans opened a factory for making gas-heated incubators. Petaluma became the chicken capital of California, and the opinions of its chicken-keepers spread far and wide through the *Petaluma Poultry Journal.*

Selective breeding also played a part in engineering the birds that would be factory farmed—the excellent egg layers and the big, fast-growing meat birds. The first breakthrough was made with laying hens. It had been impossible to tell precisely which hen had laid precisely which egg until a device called a trapnet was put across nest boxes. It yielded to entering hens but not to those wanting to exit. With exact identification, prolific egg-laying lines were established.

As the nineteenth century turned into the twentieth, chicken farming was still primarily a family enterprise. And how rewarding it was! The U.S. Secretary of Agriculture estimated that the value of poultry products in 1905 was half a billion dollars. A short five years later, only corn surpassed chickens and eggs as a money maker. Betty MacDonald (1908–1958), author of *The Egg and I*, and her husband Bob are last-gasp exemplars of family chicken ranching, which they did on Washington state's Olympic Peninsula. They raised pullets for eggs and

⋩ A small, modern incubator

fattened cockerels for roasting and frying in the late 1920s and the early years of the Depression. The never-ending work plumb wore her out. But she managed to write about it with wry humor. At the end of her tale, Bob plans to buy a new chicken ranch with electricity and running water, amenities to which Betty looks forward. According to Bob, however, nothing will really change. "Which just goes to show," Betty comments, "that a man in the chicken business is not his own boss at all. The hen is the boss."

An escalation in costs occurred as feed makers pushed premium feeds. And, as this end of the seesaw went up, the other, in the form of egg prices, went down. The *Petaluma Poultry Journal* made a sagacious and chilling prediction: "The time has passed for making money with poultry by old-fashioned methods. The farmer and poultry-keepers who are making any money today are wide awake, progressive men and women, who study their business, who take advantage of the best facilities, latest improvements, modern methods, etc. In short, they keep up with the times." In other words, Mom and Pop were on their way out. Savvy marketing people, financial wizards, and CEOs were on their way in. Petaluma pioneered the first step toward large-scale chicken farming in the form of an egg-producers' cooperative that could bargain with the feed makers and stabilize prices. The corporation loomed ever closer.

A chicken in every pot and a car in every garage—so said the Republican Party during Herbert Hoover's campaign for the presidency in 1928. (Actually, apart from the car, the saying was not original. A King of France, Henry IV [1553–1610], said, "I want there to be no peasant in my realm so poor that he will not have a chicken in his pot every Sunday.") And, in the American 1920s, beginning with chickens, the Industrial Revolution steamrollered its way into farming. The time that it took for a hen to lay a clutch, brood it, and raise her chicks deducted not only months from her productive life but also money from her owner's pockets. Apart from saving time and money, industrial chicken-keeping was given a boost by the discovery in the '20s that when vitamins A and B were added to the feed, the birds no longer needed sunshine and exercise in order to grow. The new bargain was, I will care for you if you bring me the largest possible profit at the least cost. Economics dictated confining as many birds as possible into the least possible amount of space. The CAFO had reared up and cackled.

CAFO (pronounced CAY-fo or CAH-fo—you decide) is the acronym for concentrated animal feeding operation. It can be done on a small scale or a humongous one. Nowadays, CAFOs are used for raising dairy cattle, hogs, and chickens. Beef cattle are consigned to what might be called an open-air CAFO—the feedlot. Two kinds of assembly-line CAFOs are used for chickens—one for the broilers, the other for the egg layers. Tens of thousands of chicks may find residence in a single broiler battery, and more than a hundred thousand birds in a large one for the laying hens. Here we have unnatural history of the malign sort. All the large operations are gruesome, filled with stench and death. They can be thought of as concentration camps, Auschwitz and Birkenau for animals.

The laying-hen CAFO got off to its start when poultry scientists discovered that egg-laying and light were intimately connected: the more light during the day and night,

the more eggs. Layer batteries cram as many as ten birds beak-to-rump in cages. The cages are stacked, sometimes pyramidally, sometimes like the horizontal layers of a cake. Feed is delivered on a conveyor belt. The poop may be removed dry on a conveyor belt or turned into liquid by being washed into a trough. The hens, Leghorn and Rhode Island Red cockerel/Light Sussex pullet crosses specially bred to lay at the phenomenal rate of more than three hundred eggs a year, are kept only for two years at most before they hit the stew pot. Cockerels are, of course, hatched along with the pullets that will become laying hens. But

An egg CAFO »

because male chicks are useless in the egg department, they are killed in gas chambers or incinerators.

Broiler chicks are not confined in cages but run around on litter, be it straw, rice hulls, ground-up corncobs, or peanut shells. For their first two weeks of life, they have freedom of movement. But as they grow—and they grow at an astronomical rate, for they are Cornish crosses genetically engineered to plump out almost overnight—their quarters become quite crowded. But they won't live long, even if they live to maturity. A friend acquainted with two brothers who have a broiler CAFO in the country just north of my home heard from them that they patrol daily to collect dead chicks, of which there is such a stunning multitude that a backhoe must be used to bury them. The breasts of those that survive for six or seven weeks will have swelled to eating size, and they will be slaughtered. When they are gone, their waste will be scooped up and, often, spread on fields before the next batch of chicks arrives.

Large-scale industrial production of chickens and eggs poses problems that Mom and Pop never dreamed of. One, plain and simple, is the grisly perfume of death and fecal ammonia. Then, there are the flies. Waste disposal may present the biggest difficulty, for it not only can but *will* pollute groundwater and streams. Some broiler-bird CAFOs do not clean out waste until several flocks have been processed; that can take from six months up to a year. Laying-hen CAFOs have a far greater potential to create environmental problems than broiler CAFOs. The reason is that the producer needs 125,000 broilers on dry straw or peanut shells to be formally called a CAFO; it takes only 30,000 hens on a liquid system or 82,000 on dry litter. Why can fewer hens account for more pollution than many more broilers? Because the broilers' life span is numbered in weeks while the hens are allowed two years before they reach the stew pot.

In a necessary effort to improve water quality, the Environmental Protection Agency requires CAFOs that discharge liquid wastes into lagoons or holding tanks to obtain permits from the National Pollutant Discharge Elimination System. In addition to polluting natural waters and the air we breathe, factory farming poses other risks to human health, such as salmonella and exposure to avian flu. In 2010, there was an explosive countrywide outbreak of salmonella enteritidis in human beings. The culprits were two huge laying CAFOs, Wright

A broiler CAFO »

County Eggs and Hillandale Farms, and the source of the bacteria may well have been rodent droppings, for salmonella is not communicable bird to bird. The number of eggs recalled amounted to half a billion.

And just what is salmonella? The *Salmonella* species are named for Dr. Daniel Elmer Salmon (1850–1914), who earned the first doctorate in veterinary medicine ever awarded in the United States. As the founding director of the U.S. Bureau of Animal Industry, he specialized in studying bacterial disease in animals. One species of salmonella, *S. typhi*, "smoldering salmonella," is responsible for typhoid fever. The food-contaminating bacteria that bear Dr. Salmon's name are found mainly in poultry and pigs, although foods like raw vegetables and peanut butter have also sickened people. It does not inhabit the innards of a healthy person, but someone suffering from food poisoning caused by salmonella experiences a whole range of symptoms from cramps to diarrhea. The most likely victims are babies, invalids, and the elderly. And it takes only a single cell to make someone sick.

How is it transmitted from host to human being? The Federal Drug Administration gives a partial list of sources in its well-named *Bad Bug Book:* "water, soil, insects, factory surfaces, kitchen surfaces, animal feces, raw meats, raw poultry, and raw seafoods." Note that word *raw*. Eggs, too, are vectors, nor is it just the outside of the shell that may be contaminated but also the inside. An infected hen can transmit the bacteria from her ovary directly to the yolk before the shell is formed. So, washing eggs before they are eaten does not eliminate contagion because the bacteria are in the egg, not on it. An important secret to avoiding salmonellosis is to give raw foods, from eggs to shrimp, a thorough cooking.

The poultry-processing plants associated with CAFO birds, though automated from the kill to the packaging, are notoriously hard on their low-paid workers, who

Salmonella »

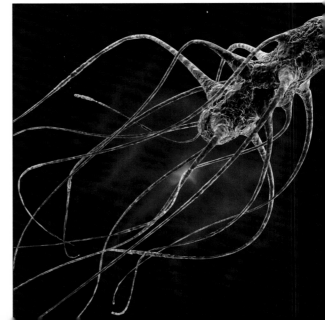

toil long hours amid blood, guts, and feathers—four billion pounds of feathers worldwide every year.

Chicken CAFOs of both kinds take their greatest toll on the birds themselves. Sickness can run rampant through birds in close confinement. Broiler chicks are vaccinated almost as soon as they emerge from the egg. We eat chickens that are filled with chemicals to keep them from getting sick. Then, hens have certain ingrained laying behaviors—wing-flapping, tail-wagging, turning around in their nests—that are impossible to perform in the confines of a cage. Equally impossible are other instinctive behaviors, like roosting, dust-bathing, and scratching. Without the ability to move, bone strength decreases and legs break. Their feathers also break. The birds may be debeaked so that they cannot peck or cannibalize their eggs or cage-mates. Boredom rules. Broiler chicks may suffocate in their own waste. I think of the little dead birds, all bones and feathers, that were part and parcel of the chicken litter that I trucked to my yard.

The best CAFOs for laying hens—if there is such a thing as *best* in super-industrialized production—have furnished cages, which are sometimes called enriched cages. With more room per hen than in conventional cages, they each provide a pad for scratching, a roosting perch, and a curtained nest box.

In an era concerned with global warming, it would be well to calculate the carbon footprints of the various kinds of CAFOs. What greenhouse gases (GHGs) are emitted, and how much of them? The EPA estimates that 6.4 percent of American GHG emissions can be attributed to agriculture. To their credit, broiler and breeder houses account for only six-tenths of 1 percent of this total, for they release significantly fewer atmospheric pollutants than cattle and swine operations. It's possible, however, to reduce the carbon footprint of the poultry industry to almost zero. Achieving this goal entails using sources of heat and cooling that do not rely on fossil fuels and managing wastes so that production of methane and nitrous oxide are minimized. The industry has a long way to go to clean up its act.

The true monstrosity of CAFOs is that they represent human divorcement from the sources of our food. Have our insatiable appetites for chickens and eggs turned us into unfeeling monsters? Can means other than CAFOs meet the demand for plump chicken breasts, meaty thighs, and succulent drumsticks? For eggs to eat and eggs to decorate at Easter?

I turn to Geri Maloney, who gave me the truckload of chicken litter, for a personal view of industrial chicken-raising. She and her husband, Kerry, have been in this business for thirty-two years, and they deal now with Tyson, which delivers chicks and fetches grown birds. "We live on a nice family-size farm," she says, smiling broadly. "I'm very pleased with being a small part of agriculture in Virginia."

And how many chicks live in their chicken house at present? "Seventy-five hundred," she says. "We raise them to six and a half pounds. Give or take a day or two, the usual growing period is fifty days. We cut back from twenty-two thousand that we brought to four pounds. They're roasters, not broilers."

The chicks are Cornish Rocks and other crosses, bred to have sturdy legs. In other words, they won't tip over and be unable to walk because of gargantuan breasts. The cockerels are

≽ The Maloneys' chicken house with four-week-old chicks

not put to death because they make just as good roasters as the pullets. "The birds have already had their immunizations when they arrive. We have a certain amount of mortality, of course, but *we* don't give them drugs. In the hot summer, I sometimes think that I'd like to live in the chicken house. Its climate is computerized, and it has cool pads—like A/C, which we don't have in our human house. The water they drink comes from our well. I think they really like it.

"When a flock goes on its final field trip to the poultry house, we have a month before the next batch of birds arrives. That's so we can keep going without hijacking the consumer. The chicken market took a big hit in the last few years because the grain needed for the birds was being turned into ethanol, not feed. Feed got very expensive."

As for waste disposal, Geri says that poultry farmers are certainly conscious of their responsibility not to pollute waterways. She and her husband have a composting shed in which dead chickens and litter sit until they're completely decomposed. The litter is cleaned out after each flock. She says, "For a while, there wasn't much demand, but now fertilizer is expensive, and farmers are clamoring for chicken litter. We have no trouble disposing of it."

She tells me that some poultry farmers are unhappy about the way that they're paid. The money earned has nothing to do with the number of birds shipped out. Rather, the farmers are ranked according to feed conversion: People who raise their birds to full weight with the least possible amount of grain get the highest marks and the most pay. Like most of the chicken farmers in our state, she and her husband belong to the Virginia Poultry Federation. "It's the voice of the poultry farmers," she says. "It helps us keep track of legislation and EPA rules that affect us."

According to its mission statement, the Virginia Poultry Federation (VPF), established in 1925 (about the time that chicken CAFOs came into being), "represents all sectors of the poultry industry, from farmers, to processors, to businesses that provide goods and services to the poultry industry." Turkeys, as well as chickens, are farmed on a large scale in the state. One of the VPF's stellar programs is a poultry-litter hotline with a toll-free number. For people who want to buy or sell litter, it lists brokers and haulers and also gives the Virginia regulations that govern such activities. Not only do the brokers and haulers need to register; the end-user must state just how and where the composted poop was used, along with an N-K-P (nitrogen-potassium-phosphorus) analysis and the location of the nearest stream or waterbody—that's the VPF's word—to the place where the litter has been spread or stored.

When I mention a few of the horror stories that rise like a miasma from the huge CAFOs, Geri says matter-of-factly, "I have never felt like we've been cruel to the birds. They live as well as we do. And I'll tell you that they played a part when our four children went to college. They didn't pay for it all, but they sure helped."

We make arrangements to get together again for lunch. I am much heartened.

Pasturing poultry is another alternative to the CAFO. This method eliminates the over-crowding and the concentration of vast amounts of manure. So, off I go to see the guru of pastured poultry, Joel Salatin. He's located at Polyface Farm in Swoope, Virginia, a short nine miles from where I live. Joel and Polyface have become nationally known through Michael Pollan's account of his apprenticeship there, *The Omnivore's Dilemma*, and through appearances in several documentary films, including *Food, Inc.* Polyface's 550 acres concentrate on grass-based animal-feeding strategies.

When I arrive on a warm, bright April morning, he's addressing a large group of people about his "living systems." A well-built man, he's wearing a cowboy hat with turned-up brim and an orange T-shirt with a University of Virginia logo; wide navy-blue suspenders hold up his jeans. It's not just his voice that I hear but also a few cock-crows and the constant egg-laying cackle of hens resident in a

≈ Pastured Rhode Island Reds at Polyface Farm

⌃ Poultry hoop house

capacious hoop house on a low hill. The hoop house is constructed something like a green-house, with fabric stretched over metal hoops.

The operations at Polyface—the farm of many faces—are extensive. Beef cattle, pigs, rabbits, turkeys, broilers, and laying hens graze the pastures. The rabbits and some of the poultry are kept in pens—tractors—that can be easily moved from one patch of grass to the next. The broilers raised here are Cornish crosses or Freedom Rangers, a buff-colored meat-bird developed in France under the Label Rouge, or Red Label, program for pastured animals. The Cornish crosses are market-ready at six to seven weeks, while the Freedom Rangers need twelve weeks. No matter the variety, Joel's broiler birds are housed in 10 x 12 x 2-foot pens enclosed with chicken wire and partially covered with corrugated aluminum roofing. On hot

« Salatin's moveable poultry pen

summer days, the backs of these pens are jacked a few inches off the ground. Each pen can hold seventy-five to one hundred broilers. Each bird has more than a foot of space.

The pasturing routine goes this way: The beef cattle graze in a field for a set period of time, eating the new grass down to a length manageable for chickens. The cattle are then moved to a new pasture, and the broilers are tractored over to the newly barbered grass. There they not only eat their greens and bugs but also batten on cow pies. Their diet is augmented with grain. To keep them safe from predators, Joel uses a guard dog to fend off attacks by coyotes, possums, skunks, and hawks.

I visit the hoop house that houses the ready-to-lay pullets—Barred Rocks all of them. They fill their quarters with the gentle murmur of hen music. I'm not the only visitor—a white goose pokes its bill through the chicken wire that stretches between the house's fabric and the ground. I see that long feeding trays are set at regular intervals from one end of the house to the other. They are narrow enough so that the pullets can't step into them and soil the feed with poop. Elevated structures between the feeding trays contain nest boxes. Next year, the ready-to-lay birds won't be Barred Rocks. They'll be black Australorps, and the year after that, Rhode Island Reds. Polyface rotates the breed of the laying hens on a three-year cycle. When their best laying days are over, it's easy to sort out which birds will be slaughtered and sold in the farm's shop. How many chickens live at Polyface? Six to seven thousand, depending on the season. Readying chickens for meat makes for a temporary decrease in their numbers. These birds aren't pets; they're business.

I sit on the porch of the shop for a while. Barn swallows make great sweeping flights through the air. Cabbage butterflies and yellow swallowtails flutter over the gravel road and the roadside grass. Car after car pulls up and, after unloading coolers, the drivers tote them into the shop and come out almost staggering under their weight.

Polyface, with a long list of steady customers, fills a niche market, catering to locavores and people who want to support sustainable green systems. More than four hundred people make regular trips to stock up on pastured poultry; one customer drives 150 miles one way. Have I bought any? No; despite their undoubted succulence, they are at least twice as pricy as supermarket chickens. But my across-the-street neighbor, who is an executive chef, brings me chicken stock that he's made with Polyface birds.

In late May, I return to Polyface with friends from West Virginia. They want to take home roasting chickens, sausage, and wieners with the skin on. The sky is a nearly cloudless blue, and the temperature is in the gentle mid-seventies. Joel, one of a team of eight people, is processing chickens in a shed open to sunshine and the air. His two elementary-age grandsons play atop the crates filled with chickens. The team stands in a row, each tending with the utmost efficiency to various phases of the operation, which will dress out more than two hundred birds this morning. The chickens are first taken out of the crates in which they've been transported to the shed. One by one, they are inserted into metal funnels called killing cones. Heads dangling from the narrow end, they are quickly killed—only their throats cut—in what would be a kosher or *halal* fashion had a rabbi or imam been standing by. Their blood drains into a tub beneath the cones. The next team member puts the chickens into scalding water, which loosens the feathers from the fat beneath the skin. Next, they are put into a large metal plucker that spins them around at high speed to remove the feathers, then spits them out after less than a minute. Their heads are pulled off; their shanks and feet, sliced off. The next step is extracting the craw and esophagus. After the birds are rinsed, they arrive at Joel. He slits the skin just above the vent. He finds the heart and liver and puts them into a pail containing water. Then, he pulls out the viscera and the lungs, which land in a five-gallon bucket on the floor. The birds are washed and rewashed as they travel along. The last team member plucks out any little feathers that may remain on the pope's nose—the bird's tail, that is. The final step is cutting the skin above the rear and tucking in the drumsticks. The blood, viscera and other innards, heads, and feet are composted. Waste not, want not.

Chicken plucker »

Joel has written a book on chicken management à la Polyface: *Pastured Poultry Profits*. In an addendum to its 2010 reprinting, he bemoans the fact that, these days, many Americans don't know how to cut up a whole chicken. "If you tell the average American," he writes, "that a breaded McNugget shaped like Dino the dinosaur is not a muscle group on a chicken, they don't believe you." But his customers began to complain that the Polyface shop did not offer cut-up chickens. Joel heaves a sigh: "A whole chicken is intimidating: it actually looks like a bird." He now carries cut-ups and sells a pound of boneless, skinless breasts for as much as a whole bird. Our laziness and ignorance are part of the pastured poultry profits.

The authors of *The Chicken Book* have lamented the events that they think signal the fall of the chicken and almost everything else—the discovery that egg-laying is affected by light and the subsequent commercialization of the egg industry in the form of monster CAFOs (though that term hadn't been thought of in their day). They saw birds—and not just birds but human resources and capital—being exploited to produce more eggs at the lowest possible cost. They state:

> Unrestrained technology, fueled by the desire for larger and larger profits, exacts a price in terms of human values that we can no longer afford to pay. There are invisible costs as well as visible ones in the destruction of nature by technology.

These sentiments were written nearly fifty years ago. Times have changed, though the CAFO still rules the roost.

CHAPTER NINE

The Resurrected Chicken

But more than pastured poultry is making inroads. In the first part of the twenty-first century, there's a downright resurrection of the chicken taking place in two forms: battery-hen rescue operations and urban chicken-keeping. In the United States, chicken-rescue sanctuaries got off the ground in the 1980s. Some of them are Animal Place, founded in 1989, Grass Valley, California; Eastern Shore Sanctuary and Education Center, Springfield, Vermont; Sunny Skies Bird and Animal Sanctuary, Warwick, New York; and Farm Sanctuary, founded in 1986, Watkins Glen, New

« A rescued hen at
Animal Place

York. Many of them care for more than CAFO rescues. In Minnesota, you'll find Chicken Run Rescue, a project of Minneapolis Animal Control, in which abused and neglected chickens are confiscated from their careless owners and given to good homes. England has several rescue efforts, including the British Hen Welfare Trust, 2005. These organizations operate for the most part on donations. But it is possible to buy hens past their prime from a CAFO or from a farm that practices pasturing. I have two sources: Joel Salatin's Polyface Farm and Sue and Jim Randall's Elk Run Farm.

The new push to keep backyard flocks of city chickens is driven in part by the new emphasis on refitting ourselves as locavores—eating as much locally grown food as we can. Urban gardening is widely acceptable. Though its clucking gets louder, urban chicken-keeping is not so accepted—at least, not yet.

Once upon a time, as recently as sixty or so years ago, it was not unusual to find small flocks in city backyards. I suspect that there were hens, if not a rooster, in my own fairly small backyard. I do know, from meeting the very old lady whose father had built my house more than a hundred years ago, that her mother had kept two cows there. But principles and practices that were common into the middle of the twentieth century have been shoved aside by the super-urbanization of America. Most of us have been separated from our rural roots. Chickens come in plastic-wrapped packages in the meat counter of the grocery store, and eggs are found, farther down in the refrigerated section, cradled in molded-plastic cartons.

Across the United States and Canada, a gallimaufry of laws, some of them draconian, dictate who may and may not raise chickens. The reasons given by the opposition cite unsanitary conditions and resultant stench, noisy cackles and crowing, and the notion that fowl constitute a public nuisance. In Homewood, Alabama, for example, it's unlawful to keep poultry within the city limits unless the birds are three hundred feet from buildings, including homes, and more than one hundred feet from the road. In Minot, North Dakota, no one can keep any kind of poultry or pigeons except for people who sell these birds in a commercial zone. Beachwood, Ohio, on the outskirts of chicken-friendly Cleveland, passed an emergency ordinance that prohibits fowl, ducks, and goats, with no grandfathering permitted. Henderson, Kentucky, outlaws not just chickens but bees on the grounds that they are a nuisance. Denver, Colorado, discourages chicken-keeping by charging fees: $50 for applying to possess chickens, $50 for a permit issued only after proof is provided that the birds' quarters will be clean and pest free, a $100 license fee, and an annual fee of $70. You must also put up two signs in your front yard for a month to alert the neighbors to your wish to keep chickens; if the neighbors object, you're out of luck.

Albany, the capital of New York, has prohibited cows, horses, ponies, donkeys, mules, pigs, goats, sheep, chickens, ducks, and geese; if you're caught keeping any of these critters, you'll pay up to $250 in fines and risk being put in the pokey for a maximum of fifteen days. Lafayette, Indiana, prohibits the keeping of any livestock within the city limits, while Boston, Massachusetts, forbids it in residential zones. Detroit, Michigan, does not allow farm animals, period. Richmond, Virginia, stipulates that you may keep chickens only if the size of your lot is at least fifty thousand square feet. Toronto, Canada, prohibits chickens outright.

Despite sanctions, some people defy the law and sneak chickens into their neighborhoods. The BackYardChickens Forum gives the story of a woman with eight birds who lives in the small town of St. Michael, Minnesota. The town specifies that a homeowner must have at least four acres to keep chickens. A neighbor informed the Powers That Be that she was breaking the law. She was told to get rid of them. So she and her husband cleared lawnmower, snowblower, and other equipment out of the big shed in their backyard. "Now," she says, "I have them in a hideout." The responses posted in the forum range from the wacky to the sensible, from telling people that the birds are rare and exotic parrots to working to change the local ordinances.

On the other hand, you can keep chickens legally in lots of other places.

A chicken-worthy shed »

Most pro-chicken cities and towns limit their numbers. Some prohibit roosters, some don't. A few forbid the keeping of gamecocks. But I was surprised to learn that Los Angeles does not place an upper limit on numbers; the city stipulates only that the birds must be at least twenty feet away from the owner's home and thirty-five feet from other residences. On the other side of the country, the health code of New York City has declared that chickens are pets; an unlimited number of hens is allowed, though no roosters, ducks, or geese.

Atlanta, Georgia, allows chickens. A fairly recent ordinance in Lawrence, Kansas, allows up to twenty hens, but no roosters. Minneapolis allows an unlimited number of hens if 80 percent of the neighbors within one hundred feet of the proposed coop agree; the birds must be kept penned. Miami allows up to fifteen hens, though no rooster; it has also decreed that chicken droppings cannot be used as fertilizer. Rather, they must be wrapped in paper and place in a closed bin for pickup by the municipal garbage haulers. In Cedar Rapids, Iowa, you may keep up to six hens; the ordinances mandate banding the birds with ID bracelets supplied by the city, which also requires would-be chicken-keepers to take a how-to class before getting any hens. Kansas City allows fifteen hens; Chapel Hill, twenty; and Santa Fe, unlimited birds. One of the morals here is: If you fancy keeping chickens, be careful about where you live.

The upswing in chicken-keeping has given rise to chicken swaps all over the country. People buy, sell, and trade poultry and eggs, some of the latter fertilized so that they can be incubated and hatched. And a new service has also hatched—chicken-sitting. The sitters will call on the birds twice a day to make sure that they are fed and watered, that the eggs are collected and cleaned, and that droppings are removed from coops. Los Angeles has Easy Acres Chicken Sitting; Just Us Hens is located in Portland, Oregon; and Green Acres is found in Denver.

Some localities, like mine, are woefully ignorant of their own ordinances. An article on urban homesteading in an April 2011 issue of my local paper stated that three hens are allowed here. One of the homeowners who was interviewed said that he'd heard that three birds are legal; he'd gleaned that information from someone at city hall. Two days later, the local television channel that carries city news declared—and still declares—that chickens are completely verboten in Staunton. You might call this tactic media intimidation: If the TV says that it's so, then it is so. Except that, according to the city code, it's not so at all.

War is being waged elsewhere, with the pro-chicken forces pitted against the anti-chicken troops like gamecocks spoiling for a fight. San Diego and San Leandro, California, have seen fierce debates between citizens and city council. People in Washington, D.C., have worked to eliminate the capital's requirement that hens be penned fifty feet from any residence. Shouting matches have also erupted in Evanston, Illinois; Portland, Maine; Albany, New York; Salem, Oregon; and Caledonia, Wisconsin. In Harrisonburg, Virginia, just twenty-five miles up the pike from me, the pros and antis have battled in the city council's cockpit. The antis prevailed. Under the new rules, four hens are allowed for a single-family home; the catch is that the home must be situated on at least two acres of land. That requirement disqualifies most properties in the city. There's an irony here: Harrisonburg is the county seat of Rockingham County, the poultry capital of the state. Its boundaries are marked by roadside pillars, each supporting a life-sized figure of a turkey.

As you can see, the New World's chicken laws are a higgledy-piggledy hodgepodge. But the thrust toward controlling our own destinies by keeping chickens is not to be stopped.

Where do you keep city chickens? In a coop, of course. And just what is a coop? The word is considered synonymous with "poultry house." Columella is most specific about the construction of a *gallinaria*, a poultry house suitable for a farm. It should

« Turkey statue, Rockingham County

face the rising sun in winter and should adjoin the kitchen of the house so that smoke, beneficial to the birds, can reach them. The house should be divided into three rooms, each with a loft for roosting, plus a ladder for reaching the roost, and each with a small window that the birds can use for access to the poultry-yard. The cells are also furnished with wicker laying baskets.

It's the word "basket" that holds significance. The Middle English *cupe* or *coupe* means "basket" (a clear reference, I think, to a nesting basket, the laying of eggs being the prime purpose of a henhouse) and came to denote, according to the *Oxford English Dictionary*, "a cage or pen of basketwork or the like for confining poultry."

The cooping used for contemporary backyard flocks is about as far removed from basketwork as is a supertanker from a coracle. Many are simply functional wooden structures, high enough for a person to walk in upright. The colors of these run-of-the-mill coops tend to be a bit more imaginative, covering the spectrum from prosaic barn-red to exotic purple and azure with silver spangles. Other coops affect a rustic elegance as miniature log cabins or A-frame

A super-fancy coop »

eyries. Still other coops rise like the proverbial phoenix from cast-off materials, from loading pallets to beat-up fifth wheels. Then there are McMansions that replicate the chicken-keepers' own house, its colors, windows, shutters, and columns, its gutters and downspouts, its door-knocker. The names that chickeners attach to these habitats on fancy wooden plaques are as varied as the architecture. They include the inevitable puns, like Cluckingham Palace and Coop de Grass. The obvious is not left out: Hen Hilton, Biddy City, and Clutch Hutch. Other times, other cultures sneak in, too: Chickenarium, Taj-MaCoop, Chez Poulet. Some people cannot resist a tad of scatology: Poop Deck and Poo Drop Inn. My favorite was found on a coop-naming discussion held on the website of the BackYardChickens Forum:

> I briefly pondered "Tweety's Farm," which is a *Chicken Run* reference. But the name I settled on was a way to thumb my nose at my jerk neighbor. He must have overheard me talking about chickens and got upset. So he stormed over and tried to intimidate me into not keeping those "dirty, smelly birds." I told him I wasn't building a chicken coop but was instead building a tool shed. I plan on mounting a plaque over the door that reads TOOL SHED.

Many cities, like Seattle, Pittsburgh, and Charlottesville, hold annual tours of backyard coops. These events are attended by the committed and the skeptical, the latter of whom go touring to confirm their suspicions of dirt, noise, and odor. They are usually out of luck.

Most backyard chicken-keepers keep clean coops. But what do you do with the chicken litter if your local ordinances do not mandate sending it out with other household garbage? High in nitrogen, it can burn tender plants if it's applied directly. The secret lies in composting the poop-laden straw. It can be put into a pile or into a 3 x 3 x 3-foot wire bin of the sort sold by gardening supply companies. When gardening season ends, the litter can be put directly on the earth or onto raised beds, where it will cool down during the winter. Come spring, no further fertilizing need be done.

Not all backyard chicken-keepers use straw, though straw's virtues are manifold. To begin with, it's inexpensive. It's easy to rake at coop- and yard-cleaning time. If it's seedy, the hens enjoy pecking at and gobbling up the grain. But other materials are also available—pine shavings that keep down odor, chopped cardboard that's more absorbent than shavings or straw, and, as a last resort, shredded paper.

The urban chicken-keeping movement is indeed a resurrection of the chicken.

CHAPTER TEN

The Conquering Chicken

"Next to the Dog, the Fowl has been the most constant attendant upon Man in his migrations and his occupations of strange lands." So writes a nineteenth-century researcher into all things gallinaceous. Found not only in their Indian and Southeast Asian homelands but also in Europe, Africa, and even Greenland, chickens were a global phenomenon, found even in South America, as the prehistoric presence of Araucanas attests. The researcher adds, "But the most mysterious, though not the most ungenial, localities in which Fowls have hitherto been found, are the islands scattered over the vast Pacific Ocean." Tasmania, Tahiti, the Marquesas, the Sandwich and Society Islands, Hawaii—Captain Cook and other European explorers found chickens everywhere.

Multicultural, multinational, and ecumenical—that's *Gallus*, the bird that has conquered the world.

The Japanese Chicken

A Japanese proverb states that it is better to be the head of a chicken than the rump of an ox. Certainly, the bird has long been an object of delight in that country.

Japan boasts about fifty native breeds of chickens. They are divided into two groups: fancy birds for hobbyists and utility birds for meat and eggs. The fancy fowl are again divided into two subgroups, those that have been in Japan for more than two thousand years and come-

White Onagadori »

lately birds that were brought in during the Heian Era (794–1192). The former are known as Japanese Old Type, and the latter, as Japanese Elegancy. Nineteen of them, including the adorable, wooly-feathered little Silkie, have been named "Natural Monuments of Japan," and one of the nineteen, the Tosa-Onagadori or Japanese Long-Tail, has been called a "Special Natural Monument." The Long-Tail cock, which does not molt, comes in various plumage colors—silver duckwing, black-breasted red, white, and buff Columbian. Its remarkable tail feathers can grow to a length of eleven or more yards. They lie on the ground behind the bird like ribbons in swirled patterns. Oh, the fancy roosters are glorious birds indeed. One breed, the Tail Dragger, sports luxuriant saddle hackles and drags its equally luxuriant tail feathers on the ground. Another, the Black Crower, is covered with iridescent black plumage. The Kagoshima Game Cock, originally bred for fighting, fans out its tail feathers when it's excited. But for all that the cocks are superbly colored and feathered, their hens are uniformly drab.

One of the Natural Monuments of the Elegancy breeds is a gamecock, the Shamo, which originated in Thailand when it was still called Siam. Shamo is a nipponization of Siam. He's a long-legged, tall, muscular bird with iridescent black, rusty, or wheaten feathers, and he stands bolt upright. Of all things, the Shamo cock defies the rules of roosterdom by being monogamous. Because he is not given to raucous crowing, it's possible to keep a breeding pair and raise Shamo chicks in close proximity to the neighbors. It's the Ko-Shamo—the bantam Shamo—that's used for cockfighting, which is legal in Japan and everywhere else in the Far and Near East. But beware: If another cock is introduced into the household of any Shamo, large or small, a fight to the death will ensue.

It was not until the late 1800s and early 1900s that the utility breeds were truly established. To say this is not to say that chickens were excluded from the menu in earlier times. A Japanese correspondent tells me that even back in the Edo Period (1603–1867), when Japanese were vegetarians for religious reasons, they were allowed to eat animals that did not have four legs. A favorite chicken dish, especially among men, has been and is yakitori—grilled chicken on a skewer. But he says, "Eggs in olden times were very expensive. They were treated like medicinal food for individuals weakened from illnesses. As everything else became expensive, price of eggs remained fairly stable, and now they are inexpensive." He remembers visiting his grandmother, who kept chickens in the backyard. She would serve him a favorite Japanese breakfast: rice topped with a freshly laid raw egg seasoned with soy sauce.

The number of utility birds in Japan is currently very small, although chickens are raised on a commercial basis, some of them in close confinement. But my correspondent tells me that free-range chickens are considered more desirable for meat and eggs. "We even eat some fighting cocks from the northern area." He hasn't tried one himself, for they're said to be rather tough and fit only for stewing.

A Japanese scholar who studies chicken genetics says, "People mostly rear them to enjoy their beautiful figure." He thinks that the fancy fowl comprise a genetic pool that can be drawn on to produce tastier meat than the utility birds provide. He cites delicious crosses that have resulted from mating Natural Monuments to Plymouth Rock, Barred Rock, and Rhode Island Red.

In the early 1980s, Colonel Sanders brought his Kentucky Fried Chicken to Japan, where the

Colonel Sanders »

fast-food chain achieved great popularity. An American acquaintance tells me that when she visited Japan, she went to one of the Colonel's restaurants. She was astonished to be served a whole chicken. "But it was tiny," she says. "Like a bantam." It makes sense for the researcher to improve not just the tastiness but also the size of utility birds. On the other hand, it's hard to eat a pet.

Chickens also figure in religious rituals and customs. Here's a sampling.

The Jewish Chicken

Archaeology reveals that chickens were raised in Israel from late Biblical times on. The birds may have come from strains found in Persia or India. The Hebrew word for a rooster is *tarnegol*, a term derived from a Sumerian word meaning "king bird." Nowadays, the Hebrew word for a rooster is *gever*, a term that happens to mean "rooster" in Aramaic.

In the early days, rabbis disagreed strenuously about whether chicken meat was or was not kosher. In Jerusalem, the bird was considered unclean, but in Galilee, it was pronounced ritually clean and, therefore, able to be eaten. While the Holy Temple stood in Jerusalem, chickens were among the animals sacrificed. After the destruction of the Second Temple in AD 70, the slaughter of all animals was prohibited. Both the feathered and the four-footed ceased to be offered to God.

Which came first, the chicken or the egg? The Talmud provides an answer to this famous question: Everything in the world was brought into being in its full perfection. Therefore, the chicken has primacy.

It's not surprising to find that chickens figured—and figure to this day—in several Jewish rituals. Hens and their eggs have long been considered symbols of fertility, and roosters, symbols of pluck and highly successful procreation. The Talmud speaks of learning kindness toward one's mate by observing the rooster, who summons his hens when he finds delectable tidbits. Once upon a time, not long ago, when a bridegroom and his bride were being escorted to the ceremony, a cock and a hen were carried before them. The message is clear: Be fruitful and multiply. Although this chicken-toting is no longer practiced, chickens, along with fish, are still featured at weddings as a primary food for the newlyweds, their parents,

Seder plate with *beitzah* »

and their guests. The reason? Both birds and fish, being blessed with abundant eggs, are potent fertility symbols. To this day, a roasted egg is set upon the Seder plate on the first day of Passover, the holiday that celebrates the Jewish exodus from enslavement in Egypt. Called a *beitzah*, the egg is one of six ritual foods and symbolizes the festival sacrifice that was offered at the Temple in Jerusalem.

A far-less-merry ritual is carried out in the pre-dawn hours of the day before Yom Kippur. It's a ceremony of many transliterated spellings—*kaparos, kapporot, kapporah,* and more, all of which mean "atonement." Orthodox Jewish communities have long practiced it and still do in many places, including New York City. The ritual goes this way: A chicken—cock for a man, hen for a woman—is held in the hands while selections from canonical psalms are recited.

Then the bird is grasped by both wings and swung very gently three times over the head while the swinger says, "This is instead of me. This is an offering on my account. This rooster [or hen] shall go to his [or her] death. And may I enter a long and healthy life." The person's sins are transferred to the bird, which is then given to the *shochet*, the ritual butcher. It is incumbent upon him to cause the bird no misery by cutting its throat quickly with one incision from a very sharp knife. After the bird is hung to drain its blood, it is donated to charity.

« A bird about to be swung

Kaparos has not unnaturally evoked strong castigation, even among Orthodox Jews. But tests have shown that chickens slaughtered in the *shochet*'s fashion do not endure suffering, that they do indeed die mercifully. The crux is, should they die at all in such a ceremony that seems a hangover from paganism? Never fear. Modern Jews have come up with a pleasingly Orthodox solution: Wave money overhead instead of a chicken, then give it to charity.

The Muslim Chicken

For Muslims, the chicken is *halal*—a permissible food according to Islamic religious law. When a chicken is killed, the butcher first invokes the name of Allah. Then, his method of slaughter is like that of the Jewish *shochet*—using a sharp knife to cut the throat, esophagus, carotid artery, and jugular vein, but not the bird's backbone. The object is twofold: to kill the bird mercifully and to drain its blood. In this day of factory farms and factory processing, concerned Muslims have asked if such processing is *halal*. The answer is yes, if God has first been invoked and the chickens are killed with a machine that swiftly cuts their throats in the prescribed manner so that all of the bird's blood leaves its body.

The problem here is that chickens killed wholesale are usually stunned first with an electrical shock. Even if Allah has been called upon before that happens, to behead the chickens completely is not *halal*.

It's easy to see how Kentucky Fried Chicken and McDonald's have found themselves objects of censure and outright scandal. Britain and France have both known Muslim outrage at their claims that their poultry products comply with religious law. French KFC outlets were purchasing their chickens from France's biggest poultry producer. But not much in the way of lost sales resulted; the evidence was not enough to stop mouths from watering when they contemplated fried chicken. My American friend who ate a whole bantam in a Japanese KFC tells me of her experiences in Java. When KFC ventured there, it was not well-received at first, mainly on the grounds that the food it sold was not *halal*. The problem was solved when KFC outlets not only hired ritual slaughterers but also posted signs that showed the face of a cleric and his guarantee that the meat was religiously fit to eat.

The Ominous Chicken

The Greek and Roman forebears of the Western world were hardly the only people to find magical powers in chickens, to read their actions, eggs, or entrails for portents, or to see hens as avatars of boundless fertility and cocks as randy symbols of ceaseless sexual energy.

Chickens have been ominous—full of omens—for millennia. Anthropological findings around the globe in the last hundred years have discovered a mort of tribal beliefs and customs associated with the birds and their eggs. It's safe to say that these modern findings reflect thoughts and practices immemorial. The ancient bargain lasts in some places to this day: I will care for you so that you tell me how to behave or give me predictions of things to come. Here's a sampling:

- The most ominous chicken in Christendom crows in the Bible's Book of Matthew. At the Last Supper, Jesus predicts that the apostle Peter will deny him three times before the cock crows. Accused by the Pharisees of contesting their leadership and put under arrest, Jesus appears shortly thereafter before the high priests and the elders, who spit in his face and hit him with open hands. Three times the disciple Peter is asked if he is an adherent of Jesus's teachings. In the words of the King James Version, the third time that he was asked, he "began to curse and to swear, saying, I know not the man. And immediately the cock crew. And Peter remembered the word of Jesus, which said unto him, Before the cock crow, thou shalt deny me thrice."

- Tibetan lamas may not eat chicken flesh because the birds eat worms and are therefore unclean.

- In southern China and Southeast Asia, tribal peoples use chickens for divination in the following fashion: Bamboo splinters are inserted in the air sacs in chicken bones. (Air makes the bones lighter so that the birds can fly.) The angle at which the splinters project is read to reveal the future.

- A bargain that Hindus in Indonesia make with chickens is, I will care for you so that you may defend me from evil. In a cremation ceremony, a chicken is tethered nearby by its leg so that evil spirits will enter the bird rather than any member of the deceased's family.

Kali »

- A tribe in the Sudan poisoned chickens in order to foretell things to come. The poisoned bird would be asked a question. If it died, the answer was yes, but if it lived—no.

- In Nepal to this day, chickens are sacrificed to six-armed Kali, Shiva's consort, who is not satisfied until her statues are drenched with blood.

- Santeria is a religion that arose from the slave trade in Cuba. It combines elements from West-African Yoruba beliefs and Roman Catholicism. Its supreme, immortal god is surrounded by *orishas*, which are elemental spirits but nonetheless mortal. People who practice Santeria believe that, with the proper observances, the *orishas* can help them to achieve the destiny for which they were born. It's incumbent upon worshipers to keep the *orishas* alive by feeding them. Thus, animals of many sorts, including chickens, are sacrificed. They must be ritually killed with one blow. The birds are not wasted but cooked up and eaten. People who did not share this belief in animal sacrifice brought suit in Hialeah, Florida, to stop the practice. The case went all the way to the Supreme Court of the United States, which ruled in 1993 that this practice, as part of a religion, was protected by the First Amendment.

- To this day, the Western world uses the furcula—the forked bone between the neck and breast of a chicken—as a wishbone. The custom may be as old as the Etruscans, who lived in the regions now called Tuscany and Umbria before the Romans arrived. They considered chickens sacred and used them for divination. When a chicken

« Wishbone

died, its furcula would be allowed to dry. Touching it would bring good fortune. The Romans adopted the custom, if not Etruscan modes of divination—after all, they had their own. It seems that people grabbing good-luck bones would break them. Depending on where you lived, either the larger piece or the smaller was deemed to confer luck and make your wish come true. The Romans brought the wishbone-breaking custom to England, where the furcula was called the merrythought because of its association with good fortune. And the English brought wishbone-breaking to the New World.

CHAPTER ELEVEN

Eggs

Ah, the egg! The magical, mystical egg! Creation myths make much of it. Sanskrit scriptures mention Brahmanda—the Cosmic Egg—a concept that speaks of an expanding universe. Today's cosmologists echo this belief all unaware when they propose the Big Bang expansion of the universe from a single highly compressed mass. The Rig Veda, a collection of hymns composed between 1200 and 900 BC, contains one hymn that also refers to the beginning of the universe: Hiranyagarbha—"golden fetus" or "golden womb"—floated in cosmic emptiness until it broke into two halves: Heaven and Earth. Chinese myth gives us a similar view: When the god Pangu, born inside an egg, broke it open, the rounded end became the sky while the lower half became the earth.

Once upon a time called the Dreamtime, Australian Aborigines were given their ways of life. They believed that the earth and its oceans, the sun, moon, and stars were all created from the contents of an egg—an egg that happened to have been found by a hungry fisherman. He made a fire to cook the egg, but as soon as he'd set it in the coals, a storm came, inundating everything with rain and more rain. Then the egg cracked open, releasing more water, along with mountains, the rainbow, and the heavenly bodies. And it washed the Dreamtime away.

Traditions involving eggs are found worldwide. In Korea, legend has it that the country's first king came from a red egg left by a flying horse. In former times, Ethiopian women who were caught eating eggs were enslaved. A Congolese tribe believed that a woman who ate an omelet would go insane. The Sema Naga of northeastern India and northwestern Myanmar proscribe the eating by women of eggs laid hither and yon, rather than in a nest, by one hen. Why? Because eating eggs laid by such a carefree hen would lead to promiscuity. On the other hand, if a woman in northeastern India offered an egg to a man, she was proposing marriage. And because eggs are potent symbols of fecundity, in the early days, German farmers coated their plows with egg

⌃ Easter eggs

whites and yolks to ensure the fertility of their fields. In China—who knows how long ago?—the custom arose of giving a red egg as a birthday gift to a child; to this day, red symbolizes happiness and longevity.

Then, we have the Easter egg, the egg that symbolizes emergent life and resurrection. Painting eggs certainly predates Christianity. Zoroastrian new year celebrations using decorated eggs date back 2,500 years. And for millennia, the Jewish Passover Seder has featured hard-cooked eggs. For Christians, eggs were long a forbidden food during Lent; the dawning of Easter Day meant indulgence at last. (I remember the Easter egg hunts that we conducted on the significant Sunday. My children erupted into the backyard with baskets and set about seeking the eggs most cleverly hidden by their father. I sat in the kitchen and waited, getting hungrier by the minute, for the eggs to come in so that they could be properly shelled and deviled for the festive breakfast.)

Just as the Western world has incorporated ancient solstice celebrations in the form of a lighted Christmas tree, I am sure that some of the Easter customs are fairly modern inclusions of pre-Christian traditions. One would be egg-rolling, a widespread activity that occurs annually in high venues and low, including the White House lawn. Another is egg-jarping, also called egg-tapping or egg-dumping, in which each participant is given a hard-cooked

« Egg dyed red

egg. A niece with an Armenian heritage calls this game "egg-fighting," which she engaged in with eggs dyed onion-skin red by her grandmother. The object of the game is to tap the small end of your egg against the small end of someone else's. The winner is the person who holds the last uncracked egg. In some contests, the losers get to eat their eggs; in others, it is up to the person who cracked each egg to eat them all (sounds a bit like indigestion to me).

Superstitions accrue to eggs as iron filings to a magnet. One that dates to the Middle Ages is that the shell of an egg must be crushed lest a witch turn it into a boat, sail out to sea, and stir up howling storms. An egg laid by a white hen in a new nest on Easter Sunday cures headaches and stomachaches; broken in a vineyard, it protects the vines from hail and, broken in a field, it keeps frost at bay. Eggshells should never be burned lest the hens cease to lay (hungry hens will eat their eggs happily, shells and all). Eggs have served as tools of divination, sometimes by being dropped, sometimes by being deliberately broken, so that the contents can be read. As for eggs with double yolks, they contain contrary prognostications, one to suit optimists, the other for pessimists: Either good luck will ensue or there will be a death in the family.

Mortal imagination has always been enamored of eggs and their close-held possibilities. The egg, that perfect form, has been recreated in jewels and precious metals. The gloriously jeweled egg was brought to perfection by Carl Fabergé, a Frenchman whose father moved to St. Petersburg and opened a jewelry store. The Russian royal family, Nicholas II and his empress Alexandra, became the younger Fabergé's patrons. To commemorate the coronation, Alexandra commissioned an egg that contained a model of the cathedral in which the emperor had been crowned. The two clocks on the cathedral tower keep accurate time. Then, to mark the completion of the Trans-Siberian railway, Fabergé crafted a huge egg that held a working model of the train.

Fabergé egg »

I can't overlook the tale of Columbus's egg. Though the story first appeared in a book published in Italy in 1565 and lies in the realm of the apocryphal, it's worth repeating. Columbus was confronted by a group of Spanish nobles, one of whom told him that if he had not discovered the Indies, then a well-educated Spaniard would certainly have set off on such an adventure and made the very same findings. Columbus did not respond to this remark but asked for an egg. When he had it in his hand, he challenged the assembled noblemen to make the egg stand upright on its end. All tried, but none succeeded. Columbus took the egg, tapped it so that it broke slightly, and stood the egg on its end. The moral of the story is that once a task has been accomplished, anyone knows how to do it.

We are not without literary eggs. One appears in a riddle in J. R. R. Tolkien's book *The Hobbit*: "A box without hinges, key, or lid, / Yet golden treasure within is hid." Eggs also figure in Book I of *Gulliver's Travels* by Jonathan Swift. Captain Lemuel Gulliver is told that the kingdoms of Lilliput and Blefescu have been waging war for the last six and thirty moons. Why?

[The war] began upon the following occasion. It is allowed on all hands, that the primitive way of breaking eggs before we eat them, was upon the larger end: but his present Majesty's grandfather, while he was a boy, going to eat an egg, and breaking it according to the ancient practice, happened to cut one of his fingers. Whereupon the Emperor his father published an edict, commanding all his subjects, upon great penalties, to break the smaller end of their eggs. The people so highly resented this law, that our histories tell us there have been six rebellions raised on that account; wherein one emperor lost his life, and another his crown.

Swift tells us that eleven thousand Big-Endians have died rather than break their eggs at the small end. Big-Endians, who live on the island of Blefescu, and the Little-Endians of Lilliput serve as satiric stand-ins for Catholics and Protestants, respectively; the thirty-six moons of war represent the conflict between Rome and England and suggest the English Civil Wars of the 1640s when Charles I was beheaded and James II sent into exile.

In classical times, though eggs were certainly eaten—and with relish—Varro, Pliny, and Columella placed emphasis on the egg as the source of more chickens—more chickens for people to eat and more cocks to scrap with one another for people to bet on. The ancient world had its share of egg superstitions and customs, like that of incubating an egg in one's bosom to assure the birth of a male child. Eggs were carried in religious processions, such as that of Ceres, and used to mark the first lap of chariot races.

The Reverend Mr. Dixon has, of course, much to say about eggs. Experimentation proved to him that the classical notion that eggs could be sexed between pullets and cockerels by their shape was not just wrong, but also idiotic. He quotes Cicero to the effect that the Roman proverb—all eggs are alike—was untrue because the inhabitants of Delos could tell just by looking at an egg which hen had laid it. He comments:

> Eggs are popularly supposed to be so much alike, that what can be said about one Egg is thought applicable to every other laid by the same species of bird—the common Hen for example; but there is nearly as much distinguishable difference between the units in every egg-basket which is carried to market, as there is between the faces in a crowd of men, or the hounds in a pack, to every hen belongs an indi-

vidual peculiarity in the form, color, and size of the Egg she lays, which never changes during her lifetime, so long as she remains in health, and which is as well known to those who are in the habit of taking her produce, as the handwriting of their dearest acquaintances.

A whole chapter in his book deals with the need in a day without refrigeration of preserving eggs for culinary purposes by various means, including immersing them in limewater and coating them with varnish or melted tallow. Anyone, however, who wants fresh eggs in the winter would do well to put in a few well-fed, spring-hatched pullets that will be ready to lay come cold weather. "Eggs," he states, "are the superfluity of the animal's nutrition."

But these days, an egg is not just an egg. There is a bewilderment of eggs to choose from, starting with size—small, medium, large, extra-large, and jumbo. What color are they—brown or white, blue or olive green? Were the eggs laid by free-range chickens? My cardboard egg cartons tell me variously that the eggs that they contained were laid by "grain-fed hens," "vegetarian hens," or "free-roaming nesting hens." Or were the eggs produced under factory conditions? If so, were the chickens fed organic mash? And what about the nutritional quality of the eggs? As it happens, what hens eat affects the quality of their eggs. Eggs can be patented if, for example, they are engineered to be consistently high in vitamin E and low in cholesterol.

How are eggs formed anyway? The amazing process takes just a little over a day. It begins with ovulation, the release of a yolk-filled ovum from a ruptured follicle in a hen's left ovary. (She has a right ovary as well, but it does not function. The reason is that an egg is so large that it needs all the room it can get.) The yolk, however, does not originate in the ovum. Its materials are rather synthesized in the liver, whence they are released into the blood. The follicle gathers in these soluble yolk molecules and inserts them into the ovum, where they become insoluble. After erupting from the follicle, the ovum is received by the ostium—the funnel-shaped opening of the oviduct—and then proceeds to travel down the oviduct's S-curves. Its first stop is the infundibulum, where the first albumen—egg white—is deposited and the egg is fertilized if the hen has mated. From there, it goes to the magnum, which is made up of circular muscles that push the egg along with peristaltic contractions. In this place, the egg, positioned small end first, collects most of its albumen. Then

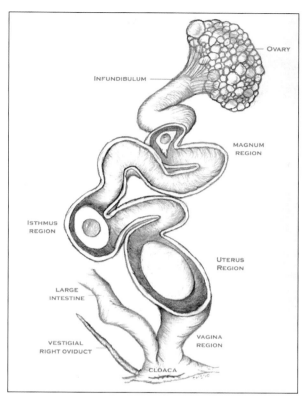

⌃ Oviduct with egg about to be laid

the egg rests for an hour or so in the isthmus, the part of the oviduct in which the inner and outer shell membranes are laid down. Next stop, the uterus (also called the shell gland), where the egg rests for up to twenty-six hours. Here it receives salt and water before the hard calcite shell is formed (calcite is a crystalline form of calcium carbonate) and here, in the last five hours, the shell receives its color.

A word here about egg color: You can't tell the color of her egg from looking at a hen's feathers. The snowy Rhode Island White chicken lays a brown egg, while the Andalusian, which may be black, blue, or splashed with white, lays a white egg. The clue to egg color is the hen's ear: white ear, white egg; ear of another color, colored egg.

When the egg moves from the uterus into the vagina, it's only minutes away from being laid. It descends into the cloaca, where the oviduct joins the digestive-excretory tract. Finally, the egg, coated with a thin covering of mucus, pops out, large end first. The mucus, formally known as the "bloom," dries and helps to protect the egg from bacterial invasion.

All of this activity is governed by hormones. Like female mammals, hens produce estrogen, progesterone, and a little testosterone. It's the last that directs hens to sport combs, which are a masculine characteristic. More than that, pituitary hormones control the production of eggs. One, known as the follicle-stimulating hormone (FSH) provokes the growth and maturation of the follicle. The other, luteinizing hormone (LH), also helps the follicle to develop and, more important, induces ovulation. The pituitary is a master gland that changes its functions with the season. Spring and summer are boom times for laying eggs. But when daylight lessens as the world turns toward its winter orbit, the pituitary cuts back on its production of FSH and LH. The hens molt, shedding all their feathers and growing fresh new ones, while egg laying lessens along with the length of the day. In their own way, hens make hay while the sun shines.

These hormones are found in a throng of creatures, from salamanders and hens to sows, mares, cows, and people. Their widespread existence is one more proof that nature is conservative. When workable paths are found, they are not discarded but rather used wholesale.

An egg is as full of furnishings as an overstuffed Victorian parlor. Within the shell, you find not just the yolk and albumen but also a host of other structures. Immediately under the shell is the outer shell membrane, which peels off a hard-cooked egg like limp paper. The air sac at the egg's larger end is found within this outer membrane. Beneath it lies a layer of thin albumen held in place by the inner shell membrane. This layer is separated from the thick albumen surrounding the yolk by another membrane. The yolk is kept centered by two

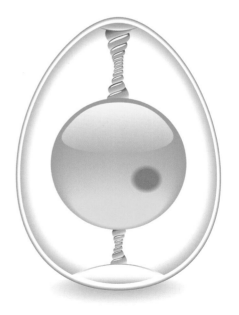

The furnishings of an egg »

chalazae, twisted membranous strings attached to opposite sides of the yolk and then, respectively, to the large and small ends of the egg. The vitelline membrane surrounds the yolk, which itself consists of yellow and white layers. (*Vitellus* is the Latin word for "yolk.") You don't see the white because the yellow is dominant. Inside the yolk is the geminal disc, also called the blastodisc. If the egg is fertilized, this disc will become a chick.

A fresh-laid egg that's been boiled is almost impossible to peel cleanly. Bits and chunks of white stick to the shell. But if the egg is aged for three or four weeks, the shell zips right off. You can tell the age of an egg by putting it in a saucepan and covering it with water. The passage of time increases the amount of air in the air sac. So, fresh eggs lie flat on the bottom and eggs with a little age tilt upward, but old eggs assume a perpendicular position.

Sometimes, a hen will lay an egg that's all albumen and no yolk. About the size of a big marble, these are called wind eggs or dwarf eggs. It was anciently thought that they were laid by a rooster, a belief that explains but also misapprehends. They are formed when a bit of reproductive tissue breaks loose and stimulates the oviduct to go through egg production. These wind eggs are completely edible, especially if they are added to a full egg and scrambled or added as binding to a meatloaf or crabcake.

Normal egg or wind egg, its expulsion into the world is announced with a cackle, the loudest sound that a hen can make, outside of the outraged squawk of a hen shoved from her place on a perch or a broody hen removed from her nest. Ulisse Aldrovandi has much to say about the cackle. Ambrosius of Nola (1457–1525) was a cleric who, among other things, translated Aristotle.

> The hen announces the laying of her egg both before and after the event, but if she is prevented from doing this, when left to herself, she breaks off her song. Columella appears to have called this song a sobbing when he says, "Hens about to lay indicate that fact by a frequent sobbing interrupted by a shrill cry." There are people who think hens grieve when they have laid an egg. Ambrosius of Nola asks why hens alone cackle or sing when they have laid an egg: "Perhaps they are especially pained not because the egg in passing out has injured them but because the place made empty when the egg has been laid has received cold air."

Aldrovandi raises an eyebrow at the notion that hens grieve and that they are penetrated by cold air. None other than Aristotle has told him that hens lay without pain. Here's another fact: The cackle registers at sixty-five decibels, far fewer than those generated by a barking dog.

In a trio of stories, P. G. Wodehouse tells the tale of Archibald Mulliner, a man with a unique talent: He can mimic perfectly the cackle of a hen that's just laid an egg, and it is with that precise talent that he woos his love.

Eggs that have been fertilized are perfectly edible. Embryos do not develop until eggs are incubated. After you collect new-laid eggs, they usually need cleaning to rid them of a bit of fecal material or, in the worst case, the yolk and white of a broken egg. Cleaning is best accomplished not with water but with a soft cloth. If a stain is stubborn, it may be wiped off with a damp sponge. Using water is not a good idea because it removes the bloom and can allow bacteria to enter the porous eggshell. Eggs should be stored small-end-down so that the yolk stays in place.

Imagination revels in contemplating eggs. I remember looking inside the windowed sugar eggs brought—laid, for all a child knows—by the Easter rabbit. I gazed with pleasure at the cottages, castles, and pastel fairyland snuggled embryonically inside. Imagination has pitied the hapless Humpty Dumpty and wishfully, greedily dreamed of the goose that lays golden eggs. It has transformed a deity into a swan that treaded Leda, earthly queen, who laid an egg with triple yolks: Castor, Pollux, and Helen of Troy. And it gazes through the eyes of Frances Hodgson Burnett at the pair of birds building a nest in *The Secret Garden*: "In the garden there was nothing which was not quite like themselves—nothing which did not understand

A sugar egg »

the wonderfulness of what was happening to them—the immense, tender, terrible, heart-breaking beauty and solemnity of Eggs."

Most eggs are tiny, be they the eggs of a fish, a cow, or a human being. But the hen's egg is enormous. The authors of *The Chicken Book* give this perspective to its great size: "The ova needed to produce many thousands of elephants would fit comfortably within the shell of even a small hen's egg." Just why are eggs so big? It has to do with the dinosaurian ancestry of the chicken. The closed-up egg, a lovely self-contained unit with a tough, leathery shell, originated with the reptiles and was subsequently inherited by their progeny and their evolutionary offshoots. One of evolution's miracles was hardening the shell of the chicken's egg.

In the natural scheme of things, the point of an egg is the production of a chick. A hen goes broody. She lays a clutch of eggs, that is, and proceeds to incubate them. Twenty-one days after the clutch has been completed, if the eggs have developed normally, they hatch.

⩘ Chick development within the egg

And, though he was off, saying that less time was needed for incubation in summer than in winter, the course of incubation is just as Aristotle described it more than two thousand years ago. After the egg is fertilized in the infundibulum, growth starts immediately in the germinal disc. If you crack a newly laid fertile egg, you can see the disc, which is a tiny translucent spot near the middle of the yolk.

The egg's first twenty-four hours of incubation are chock-full of activity. As the germinal disc grows, the alimentary tract appears, followed by the vertebral column, a rudimentary nervous system, the head, and the eye. The second day sees the development of the ear and a beating heart. The nose, legs, and wings begin to appear on the third day. Day four brings the beginning of a tongue. Day five is hugely important: Formation of the reproductive organs gets underway, and the embryo's sex is differentiated. Days six through twelve see lots of beginnings—beak, feathers, and hardening of the beak. Scales and claws appear on day thirteen. On day fourteen, the developing embryo moves into the proper position for hatching. On day sixteen, the beak, scales, and claws become horny. The embryo turns its beak toward the egg's air sac on day seventeen. On day nineteen, the yolk sac begins to enter the embryo's body cavity; it will act as nourishment till the chick learns to fend for itself. On day twenty, the yolk sac is drawn completely into the body cavity. The embryo fills the egg except for the air sac. The chick, peeping and downy, hatches on day twenty-one.

The chick grows under the warmth of a mother, be she a hen or a brooder box with a light. At six weeks or so, the cockerels differentiate themselves from the pullets by growing notice-ably large combs and the beginnings of spurs. And by that time, both sexes are clad in gawky, quill-like pinfeathers. The chicks cared for by their natural mothers learn early on to fend for themselves. As for brooder birds, when they are fully feathered, at twelve to sixteen weeks, they can be moved from brooder to coop. At about eighteen to twenty weeks, cockerels discover their salacity, and pullets begin to lay. At a year, cockerels become roosters, and pullets, hens. The beauty and solemnity never cease.

CHAPTER TWELVE

The Scientific Chicken

A scientific research tool nonpareil—that's the chicken. As the National Institute for Health has said, "The domesticated chicken is the premier non-mammalian research model organism." The birds have long served as laboratory animals for the study of a host of biological puzzles. It was through observation of the hearts of embryonic chicks that William Harvey (1578–1657) discovered the workings of the circulatory system and so overturned the long-held medieval notion that the liver was the source of blood. Centuries later, Louis Pasteur (1822–1895) used chickens in his work with immunology. In a project designed to investigate chicken cholera, a flock-decimating disease, he made the kind of discovery that happens only through serendipity. Before he went on vacation, he left a batch of full-strength cholera bacteria with his assistant, to whom he gave instructions to administer the stuff to the experimental flock. But the assistant, clearly someone who believed that while the cat's away, the mouse should play, decided to go on vacation, too. On Pasteur's return, the stuff was still sitting there in an attenuated condition. Nonetheless, he gave it to the flock, which promptly sickened, only to make a quick recovery. Nor could the birds be reinfected with cholera. Aha—immunity! And this discovery led to the formulation of other

Louis Pasteur »

inoculants, one of which immunized cattle against anthrax. Pasteur came up with the name "vaccine" to honor Edward Jenner (1749–1823), the British scientist who had discovered that inoculation with cowpox could immunize people against the disfiguring ravages, the sheer deadliness, of smallpox. Both Jenner and Pasteur were, of course, learned in Latin, a language that calls a cow *vacca*.

The language of genetics—genetic linkage, alleles, epistasis, and more—comes from work done on chickens more than a century ago by various researchers, who were interested in such matters as what makes a red feather red and a comb large and floppy or small and neat. The chicken was the subject of the earliest genetic mapping.

Because it is easy to get at and manipulate embryonic chicks, they have also figured in biomedical research dealing with development, viruses, cancer, and evolution. The bird has provided models for studies on aging, for studies of developmental disorders like cleft palate, and for studies of diseases like muscular dystrophy. Embryonic chicks have performed stellar service in determining whether prenatal exposure to chemicals like chlordane and sodium thiopental can cause abnormalities in human development. (Sodium thiopental is notoriously one of the drugs used in execution by lethal injection.)

Embryonic chicks have memory encoded in their genes. One special memory is that of the proto-bird, the archaeopteryx, which had teeth. Scientists at the University of Wisconsin-Madison and the University of Manchester in England announced in 2006 that they had succeeded in awakening that memory. Some mutant embryos possess a recessive and fatal trait called talpid2. During incubation, they begin to grow teeth just like those found in ancestral proto-birds. These mutants do not survive long enough to hatch. But normal embryos have latent talpid2, and when the gene was activated, it stimulated embryonic jaw tissues to form teeth just like those of their ancient ancestors. No grafting or transplants from a mutant embryo were involved. The developmental biologist in charge of the project says, "These results provide clear evidence that these chickens possess the memory of the past; they have retained the ability to make teeth under certain conditions. What I am describing is evolution."

In the late nineteenth century, the birds were used as experimental subjects in order to distinguish between germ cells and somatic cells. After biologists had learned that a body comprised cells, an argument arose: Were the cells identical to the male reproductive cells that had fertilized the egg? The answer—there are two kinds of cells—came from working with chicken embryos. The part of the rooster's sperm that was believed to contain reproductive cells was removed. The embryo was allowed to develop, hatch, and grow up. Behold: A bird that looked to be an everyday chicken—except that it could not reproduce. The experiment resolved the argument by providing proof that two kinds of cells are needed, those that reproduced and those that developed into body tissue.

Chickens are studied on their own behalf. They are useful birds for investigating Mendelian genetics. The huge-breasted broilers, known as Cornish crosses or Cornish Rocks, are not a breed at all but rather hybrids of meat-type chickens; they resulted from selective breeding (which is a non-aggressive form of genetic engineering). At six weeks old, they are far larger than chicks the same age from standard breeds. Poultry scientists also seek cures and means of prevention for various diseases that affect domestic fowl. They've discovered, among much else, that leukemia can be transmitted in chickens. Avian influenza is of special concern because the virus that infects birds can mutate and infect us, an infection for which we have no natural immunity.

Significant research on the growth rates of chickens has occurred at Virginia Tech since 1957, when Paul Siegel joined the faculty in the College of Agriculture and Life Sciences. A genial and now-balding man who grew up on a poultry and tobacco farm in Connecticut, he took White Plymouth Rock chickens and developed two lines, one for a high growth rate, the other for a low growth rate. At eight weeks of age, the former weigh as much as ten times more than the latter. The two lines have been used for a raft of studies. Researchers at Sweden's Uppsala University, the Massachusetts Institute of Technology, and Harvard identified a gene found in all the high-growth line but in few of the low-growth birds—a gene that seems to govern appetite. The discovery has implications for investigating weight issues in both animals and human beings. A current project of the Uppsala-MIT-Harvard team investigates animal domestication by focusing on the gene that encodes the thyroid-stimulating hormone receptor protein. The amazing point of this research is to determine if domestication, some thousands of years ago, gave rise to a mutation in the genetic makeup of chickens. If so, proving mutation by domestication would be a scientific first.

⌃ Paul Siegel's Plymouth Rocks

Another study related to growth in chickens is being conducted by Michael Denbow, a professor of animal and poultry sciences at Virginia Tech. He is investigating a most peculiar phenomenon: Some neurotransmitters work one way in mammals but the opposite way in poultry. The neurotransmitter ghrelin, for example, increases the mammalian appetite but decreases it in chickens. Obestatin works the other way around. Understanding the biological mechanisms here would lead to understanding growth in poultry and in humankind.

Dr. Siegel focuses much of his research on estimating genotypes from phenotypes. In other words, he looks at his chickens' observable characteristics and tries to extrapolate from them the birds' genetic coding. Dr. Siegel says, "The question becomes, 'Now that we have the technology, do we have the populations to support the use of the technology?' The answer is yes. Not only do we at Virginia Tech have the populations, we have the complete pedigree for fifty-three generations." These special chickens may offer insights into such disorders as anorexia nervosa and obesity, both of which are similar to difficulties suffered by high- and low-growth chicken lines. The great weight of the biggest chickens and the low food intake of the smallest have led to many health problems that are exacerbated in each new generation.

Modern scientists have come up with some really odd findings when it comes to survival of the fittest. A pair of British researchers in conjunction with Stockholm University and the Swedish University of Agricultural Sciences discovered that social status governs sex among chickens. The researchers investigated the copulatory habits of two dozen free-ranging chickens in Sweden. As we all know, roosters tend to be big and bossy. They can have their way with smaller hens as often as they want, especially if they are top roosters in the social order. It behooves hens, of course, to mate with the biggest and best in order to give their chicks a head-start in life. The study, though unscientifically small, does show that roosters can control the amount of sperm emitted during copulation. If a cock has no competition, he can conserve his sperm to inseminate every hen in his harem. Inferior roosters will try to mate with the most desirable hens, but their sperm conservation has to do with the likelihood that they will not successfully inseminate a superior hen. They save their sperm for lesser ladies. The ladies, be they upper crust or lower, have their own strategy: sperm squirting. If a hen is trodden by a rooster that she does not favor, she'll eject his sperm before he has finished copulating. Nor will he know that he has been squirted, spurted, and spurned. According to the study, it's hens that actually rule the roost.

Because hens are easy to work with and observe, they are ideal subjects for science projects performed by students in all levels of primary education from elementary school through high school. The goal of one project, for example, was to determine if hens would lay more eggs if they were

Chickens mating »

provided with artificial light to extend their day as compared to hens that had only natural light. The answer was—and is—yes. Another project was designed to determine if the student's Light Brahma bantams would prefer one color of cracked corn over another. Orange won over red, and both were far more appealing than blue or green. Still another project considered such matters as whether store-bought eggs cracked more easily than the eggs of free-range hens. Considerable mayhem was involved in finding out that the answer was no.

An appealing science project aimed to discover if hens would lay more eggs if they heard music. Here's the objective written by Hazel, a sixth grader:

> One day I was out in the chicken coop. I was listening to my iPod. All of a sudden, I thought it would be interesting and fun to test if music affects chickens' laying patterns. I talked to my partner and we both agreed, and wondered if anybody had ever tried the experiment before and thought it would be a unique project. We would like to discover if music really does affect chickens' production levels.

Hazel and her partner picked three pieces of music—a Led Zeppelin rock song, a reggae number by Bob Marley, and a section of Mozart's Fifth Symphony. Each piece was played repetitively for a week between the hours of 8:00 AM and 4:00 PM. They carefully measured the hens' food so that it remained a constant. "But," Hazel says, "we did not control their water because we do not believe in animal cruelty." Every day they recorded the weather and the number of eggs laid. In the week that Led Zeppelin was heard over and over again, the hens produced thirty-three eggs. Mozart came in second, having inspired thirty eggs. Reggae was a pitiful last with a mere twenty.

Cancer is a scary disease. It's characterized by the abnormal replication of cells. Chickens have helped to elucidate its mechanisms. They were primary players in a discovery that

Peyton Rous »

led to the 1966 award of a Nobel Prize in Physiology or Medicine. The recipient was pathologist Peyton Rous (1879–1970), who had made a breakthrough study, published in 1911, in which he explored supposedly spontaneous tumors in chickens. He found that the tumors, far from being spontaneous, were caused by a virus, now known as the Rous sarcoma virus (RSV). It was not God nor bad behavior that triggered growth in cells. Viruses could now be implicated in the development of several types of cancer. The sneaky work that they do is to damage genetic material within a cell and so create an oncogene—a cancer gene—that causes the out-of-control proliferation of cancerous cells.

When you go for your annual flu shot, you're always asked if you're allergic to eggs. The reason is that you'll be injected with a vaccine manufactured from chicken eggs. The process of impregnating eggs with the flu virus in order to make a vaccine was discovered in the early 1930s. When the United States went to war in the 1940s, the U.S. military developed a vaccine with killed viruses that proved mighty successful. Since then, technology has found ways to keep vaccines up to date for the rapidly mutating flu viruses. And it responds quickly to pandemics like swine flue. But the vaccines are still made from chicken eggs.

It's high time to look at transgenic chickens—chickens, that is, with DNA from other organisms, including some human genes, artificially introduced into the germ line. They have been created in laboratories and are now being raised on a commercial scale. These genetically modified chickens are birds that constitute what a science writer has called "a pharmaceutical bioreactor, one that can meet the growing demand for protein-based human therapeutics." One of the tricks here has been to insert human proteins into an egg that produces transgenic chickens capable of

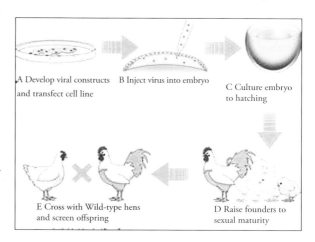

A Develop viral constructs and transfect cell line

B Inject virus into embryo

C Culture embryo to hatching

D Raise founders to sexual maturity

E Cross with Wild-type hens and screen offspring

The making of a transgenic chicken »

passing on their inserted foreign genes through ordinary reproductive methods to their chicks. Several hundred different antibodies are being tested against cancer, viral diseases, and autoimmune diseases like multiple sclerosis.

The transgenic chicken may very well be one of the best solutions to a drug-making problem. High demand for currently manufactured biopharmaceuticals—drugs produced by living organisms or cultured cells—not only clogs manufacturing facilities but can bring them to a standstill. In fact, a single blockbuster drug, like Enbrel for treating rheumatoid arthritis, can sabotage manufacturing capabilities because the company must make large capital investments for new production lines in order to come up with enough supply to satisfy clamoring doctors and their patients. New biopharmaceuticals, like vaccines and monoclonal antibodies, struggle to come on stream because existing facilities do not have the capacity needed to produce them. More and more of these drugs are under development, but conventional methods of making them have not kept pace.

Enter transgenic animals—animals genetically engineered, that is, by the insertion of a modified gene or a gene from another organism. The animals become the manufactories for drugs needed to fight infectious diseases, cancer, and a host of autoimmune diseases, like lupus, psoriasis, rheumatoid arthritis, and Crohn's disease. These living factories produce antibodies, which are the proteins naturally present in the body or produced in response to an antigen. An antigen is foreign matter—bacteria, viruses, fungi, parasites, chemicals; when an antigen enters a living creature, it stimulates the production of an antibody—a warrior designed to rout the enemy. And the enemies are legion. Many kinds of mammals have been used as bioreactors for the production of therapeutic proteins. Various human proteins that act as antibodies have been introduced into transgenic cows, sheep, goats, and rabbits which then secrete them in their milk or carry them in their blood. The problem is that enlisting the services of big transgenic mammals is a time-consuming process. Chickens are small, but they don't lactate. So, where do they come in?

Scientists studying the use of chickens as bioreactors say, "Although there are an increasing number of options for production systems for therapeutic proteins, it is recognized that the resources of commercial production create a bottleneck. Pharmaceutical proteins produced in eggs might have significant advantages for specific target drugs, lower costs than either cell culture or transgenic mammalian systems, and faster scale-up." When this statement was made in 2005, it contained some wishful thinking. Since then, a good bit of successful effort has been put into the fabrication of the transgenic *G. gallus domesticus*.

Such genetically modified birds can be used in two ways. One is to facilitate the study of vertebrate evolution. The other is to create a super-efficient bioreactor. The reasons for preferring chickens include the facts that eggs, not blood serum or milk, are easier and cheaper to deal with and that eggs can be gathered quickly in nearly unlimited quantities. The eggs are naturally rich in immunoglobulins, which are groups of structurally related proteins capable of acting as antibodies. The protein content of a yolk is 20 percent, and that of the egg white, 11 percent. Egg whites produce the pharmaceutical proteins. And those proteins, which are remarkably similar to those found in human beings, can be bioengineered to be fully human. Then, because eggs hatch in only twenty-one days, supplies of transgenic chicks are readily available. Scientists at Daegu University in Korea have lit up the world of transgenics by producing a line of chickens that expresses eggs blessed with a protein that has a glowing green fluorescence. By 2009, Louisiana State University used eggs laid by transgenic hens to produce two "biobetters," which it defines as drugs better than the originals. Its discoveries enabled a company named TransGenRx to enter the marketplace on a limited basis.

Quite a few companies have entered the chicken bioreactor business. Many have names as imaginative as TransGenRx. They include Aves Labs, Avian Immunology, AviGenics, Gallus Immunotech, Origen Therapeutics, and TranXenoGen. The last refers to its biobirds as "chimeric." In its mythical Greek incarnation, the chimera was a fire-breathing beast with a lion's head, the body of a goat, and a serpent's tail—a monstrous hodgepodge. The chimeric chicken breathes no fire and keeps on looking just as any normal chicken should. Its modified genes, a man-made hodgepodge, comprise the chimeric element. These companies focus on producing and purifying several kinds of antibodies, especially IgY, and reproducing transgenic hens that lay eggs rich in immunoglobulins. IgY stands for immunoglobulin Y, which is the major antibody found in birds and reptiles. Eggs also contain IgA and IgM. Avian Immunology sells one to four bioreactive hens for $135 each for birds raised in close quarters, $270 each for cage-free birds. The price for each bird drops slightly if more than four are ordered. The caged birds are given a foot of space each (shades of the CAFO!). Gallus Immunotech takes another tack. In a publicity release, it says:

> Making an IgY antibody isn't that different from producing a rabbit antibody, except that it's easier because bleeding the animal is not necessary. Perhaps the most difficult part of raising antibodies in chickens is to provide a suitable environment for them. Hens prefer to roost on sticks at night, lay their eggs in a nesting box and have regular "dust baths" in sand or wood shavings. At Gallus Immunotech, we provide the environment for our hens. After all, a happy, unstressed hen is likely to produce a good IgY antibody.

Now, that sounds better!

IgY can help in treating stomach ulcers; it can save the lives of third-world children suffering from otherwise fatal intestinal diseases. It also has veterinary uses, including immunizing chickens against *Salmonella*, which, if not wiped away, can pose a threat to the human consumers of meat and eggs.

Companies working on and with chicken bioreactors to make drugs face constraints that are due to the unfortunately persistent fact that federal regulatory agencies are sluggish at best to issue guidelines for approval. Nonetheless, despite the lack of prompt action, the Federal Drug Administration cannot hold back the rising waves of research, much less the making of transgenic chickens—not any more than King Canute could hold back oceanic tides.

Some of the most interesting bits of chicken science to come along in the first two decades of the twenty-first century are answers to the problem of what to do with the four billion pounds of chicken-feather waste that snows the world every year. Incinerate them? Bury them? No! Normally, it takes five to seven years for a feather to degrade. But in 2009, a young female scientist in India discovered a bacterium that can decompose the feathers in only two to four days. The U.S. Department of Agriculture's Agricultural Research Service has found an even better solution, one that not only degrades but recycles feathers. They can

⌃ Tons of chicken feathers

be converted into a plastic with properties similar to those of polyethylene and polypropylene—except that feather plastic is basically natural, not synthetic. The technology is still too expensive for the marketplace, but like the prices of computers and other high-tech devices, it should eventually become affordable.

By far the most exciting science involving chickens is the complete sequencing of their genome, a feat announced on March 1, 2004. The announcement came from the National Human Genome Research Institute, a member of the National Institutes of Health. The feat actually took place in St. Louis under the auspices of the International Chicken Genome Sequencing Consortium. Richard Wilson at the Washington University School of Medicine led the team of more than 175 scientists that assembled the genome of the red jungle fowl. The subject of their efforts was known prosaically as RJF #256. RJF stands for Red Jungle Fowl, and she was a hen, chosen because she came from a line that had been inbred for scientific purposes. This circumstance made her genome more uniform than those of chickens bred higgledy-piggledy. More than that, red jungle fowl are the wildlings from which chickens were domesticated. After she had provided samples of her DNA to the sequencing team, team member Jerry Dodgson adopted her and took her home to Michigan State University, where he serves as a professor of microbiology and molecular genetics. His official MSU website shows him tenderly cupping his special hen in both

⌃ Jerry Dodgson holding RJF #256

hands. A big beautiful bird, she was hatched in April 1995, and died in the winter of 2003 of natural causes. At the age of almost eight years, she was a very long-lived hen. Stocks of her frozen blood and DNA exist today to fill requests of those wanting to use samples from the reference sequence bird for certain experiments. About sequencing the chicken genome, Dr. Dodgson says in an e-mail, "Were we starting over today, we'd have probably chosen an inbred White Leghorn male." The reason for choosing a male over a female bird is that the sex chromosomes of roosters and hens are a mirror image of the sex chromosomes of men and women. It's women and roosters who are homozygous with XX and WW chromosomes respectively, while hens and men are heterozygous with WZ and XY respectively. Dr. Dodgson has also done some disease resistance and transgenic work on chickens. In the first decade of the twenty-first century, he was involved in the effort to map and sequence the turkey genome.

At the time of the announcement, the sequencing of the first non-mammalian vertebrate was in draft form. It has since been finalized. Not only is the chicken the first bird to be sequenced, but it is also the first agricultural animal and the first descendant of the dinosaurs to have its base pairs of sequence lined up.

And, oh my, the chicken genome has about a billion base pairs and 20,000 to 23,000 genes. That compares to about three billion base pairs and 20,000 to 25,000 genes in the human genome. (Base pairs: DNA contains molecules of adenine, thymine, guanine, and cytosine. The first two form their own pairs, as do the latter two, and in these pairs—AT and GC—the genetic information that makes a person a person and a chicken a chicken is encoded.) One difference between the chicken genome and those of the mammals—rats, mice, dogs, chimpanzees, and human beings—that have been sequenced is that the chicken genome checks in at about a third of the size of the others. Nonetheless, it's been learned that chicken genes involved in coding for eggshell-specific protein have mammalian equivalents that are involved in bone calcification. But chickens have some genes that code for light-dependent enzymes that mammals no longer possess. That they have slipped away from us reflects the fact that, unlike early mammals, we are no longer night prowlers, sheltering in the darkness from predators.

Sequencing produced some surprises about chickens. The birds have a big family of genes that affect their sense of smell. So chickens are most likely able to catch the odor of fresh kitchen scraps put in their yard, the mash in their feeders, and the air-cleansing scent of rain. But they have few genes that affect their ability to taste. Does this explain why chickens think that bugs are delicious?

What good is this knowledge? And who cares? The aims of research are agricultural and medical, notably with projects that seek to eliminate avian influenza and Marek's disease or reduce their incidence. Chicken breeders certainly care because, with genetic information, desirable traits like big meaty breasts and abundant egg production can be optimized. Sequencing will also aid in enhancing the nutritional value of chicken meat and eggs. It turns out that chickens and people share more than half of their genes, though people have significantly more DNA. (We share 88 percent of our genes with rats.) Researchers are looking for genes that bolster natural disease resistance in the birds. When those genes are found, then it will be possible to see if the same genes occur in human beings.

A medical breakthrough has been made by using chickens to study the genetic pathways of several human eye defects that cause blindness. They include retinal degeneration and sex-linked retinal dysplasia. These defects arise from genetic mutations, which can be traced and identified in chickens. With this knowledge, it becomes possible to correct genetic errors that cause eye disorders.

Chickens have a gene that was formerly thought to occur only in people; it codes for interleukin-26 (IL-26), a protein that acts in immune response. Thus, the chicken is a bird with which researchers can look into just how this protein works. And scientific exploration of the divergence of DNA in the human and chicken genomes will help to identify the distinctions between birds and mammals.

Gallus also reveals much about the basics of biology. One researcher writes: "The chicken has been used as a model in developmental biology for over 100 years." To illustrate his point, he cites the embryonic chick's limb bud, which is a structure, like the bud of a plant, that has the potential for growth and differentiation. One of three members of a family of pathways, it is a protein imaginatively known as a sonic hedgehog (SHH). The other two are the desert hedgehog and the Indian hedgehog. Whence comes "sonic hedgehog," a term that definitely holds a scientific chortle? The Hedgehog (Hh) molecule was first discovered in 1978 by Eric Wieschaus and Christiane Nusslein-Volhard, who later won a Nobel prize for this breakthrough. For a name, they seized upon hedgehog because some genetically defective fruit fly larvae were stubby and looked hairy. It was found in *Drosophila* fruit flies, where it establishes the body plan—the thorax connected to the abdomen, and so on—and affects the fly's embryogenesis and metamorphosis. When the third hedgehog was discovered, Sega's game Sonic the Hedgehog had just come out. The word "sonic" slid easily onto the new-found protein.

The same body-planning mechanism is at work in vertebrates. The SHH signaling pathway gives embryonic cells the information that they need for proper development, and it therefore plays a huge part in the genesis of vertebrate organs, among them the brain and the nervous system, the face, limbs, lungs, teeth, and hair or feathers. Genetic defects in this pathway, which are not uncommon, lead to a horrendous range of ailments from mild cleft palate to lethal cyclopia, which is a congenital disorder characterized by a skull malformation that has a single eye cavity instead of two. Jay Hirsh, a friend who is a drosophilist of world renown, writes that there is "much angst among human genet-

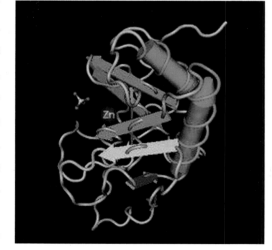

Sonic hedgehog »

Zebra finches »

icists, who hate to tell parents that their congenitally ill children are suffering from a mutated gene with a cute name." Study of the chick's limb bud, which is the expression of its SHH pathway, should lead to a greater understanding of organogenesis in human beings and, one hopes, strategies to overcome mutations.

The second bird to gain a fully sequenced genome is the colorful zebra finch, an Australian songbird with a creamy breast, a black-and-white barred back, and bright orange ears and bill. Its genome was announced in 2010. Unlike the chicken, the zebra finch learns its songs from its elders, a fact that has bearing on the ways in which human beings learn their vocalizations. How the finch learns will provide clues about how we learn.

Far more than acting as an aid to artful breeding and medical innovation, the chicken genome offers a fine tool for the study of human evolution. As Jeremy Schmutz and Jane Grimwood put it in *Nature* magazine:

> Now that the human genome sequence is essentially finished, researchers would like to do more than just identify the sequences that are translated into proteins. They also want to understand all of the regulatory structures present in a genome— structures that might, for example, adjust the amount of protein manufactured from a particular gene. These structures are collectively known as functional elements, and the chicken, having diverged from humans more than 310 million years ago, is considered the best example of an "outgroup" with which to identify them. Because enough differences between the human and chicken sequences have accumulated

over this period, one can zero in on the precise base pairs that evolution has left alone for all these years.

Nature is not wasteful. Once successful strategies have come into being, they are not thrown out in favor of newer models. Strange and marvelous to say, the chicken and the human being share a host of conserved elements—a host that amounts to some 70 million base pairs of sequences. The chicken genome will shed light on vertebrate evolution by adding details that were not available from the sequences of mammals. The consortium that unraveled the chicken genome says, "For nearly every aspect of biology, it allows us to distinguish features of mammalian biology that are derived or ancient, and it reveals examples of mammalian innovation and adaptation."

Clearly, the chicken is not just a bird that has been sequenced. It's also a bird of considerable consequence.

CHAPTER THIRTEEN

The Storied Chicken

Don't count your chickens before they've hatched.

This proverb is nearly universal. So is "Curses are like chickens; they always come home to roost." Because every part of the world has chickens, every part of the world has poultry proverbs. Here's one from China: "If you're going to have a roast, a chicken is better than a phoenix." From Germany: "The chickens don't mourn when the chicken-keeper dies." From Russia: "A sleeping fox counts chickens in his dreams." From Africa: "Grain cannot get justice in a chicken's court." And from Spain: "*La gallina de amba se caga en la gallina de abajo*—the chicken above shits on the chicken below."

And it's not just in proverbs that chickens and their eggs have claimed residence. They've found a multitude of homes in writing and speech—songs, nursery rhymes, creation myths, fables and fairy tales, TV shows, and more, though it's only their name that figures in the *Chicken Soup for the Soul* series of gooey self-improvement books. Luckily, the birds have found far more worthy roosts. Along with the fleur-de-lis, cocks are one of France's national symbols. They have served, too, as mascots for sports teams, like Peno the white cockerel for France's 1984 soccer team and the rooster Footix for France in the 1998 World Cup. The University of South Carolina's nineteen varsity teams are known as the Gamecocks, in honor of Thomas Sumter, a Revolutionary War hero known as the Carolina Gamecock because of his no-holds-barred fighting. Chickens have also been chosen as state birds, like Rhode Island's Rhode Island Red and Delaware's Blue Hen chicken. Never mind that the latter is

Footix »

not a recognized breed. Although poultry scientists have succeeded in bringing about birds with decidedly blue feathers, they don't breed true.

Nor do chickens care a whit if they figure in pop culture or more elegant venues, like myths, Chaucer, Shakespeare's *Hamlet*, and William Carlos Williams. A few of their literary appearances have already been mentioned—Aesop's fables, Pindar's ode for the boy Ergoteles who won the foot race, and Socrates's deathbed instruction that Crito sacrifice a cock to Asclepius.

Chickens certainly have hopped their way into our vernacular. Cowardly folks are chicken, chicken-hearted, or chicken-livered; they have chickened out. Pretty young girls are—or used to be—known as chicks. (Think of the downy, dyed pink and blue Easter chicks that were sold years ago.) Rooster serves as a nickname for a loudmouth. A person who's the cock of the walk is someone who thinks—not is, but thinks—that he's at the top of the human pecking order. The same goes for the cockalorum. Then there's the cockamamie cock-and-bull story. The place in which cockfights used to take place—a cockpit—is now the part of an aircraft or ship in which, far from the fray, the pilot or helmsman works. If a sum of money is chicken feed, it's not very much (apparently, chicken mash sold for much less in the olden days than it does now). Madder than a wet hen means being enraged to the nth degree. Cockeyed can be taken in two ways: schnockered—he bent his elbow until he was cockeyed—and nonsensical—that's a cockeyed notion. People can be cocky or cocksure. To be cooped up means to be trapped indoors. A person who embarks on an unsuccessful venture may be said to have laid an egg. On the other hand, an egghead—the opposite of a dumb cluck—is someone seriously interested in matters of the intellect. The interdiction against putting all your eggs in one basket warns against centering all desire on one goal; to avoid disappointment, it's best to diversify your hopes and interests. To egg someone on is to prod someone to do something and do it now. Sometimes, summoning an imaginary wishbone, we yearn for "a lucky break." We speak, too, of having our hackles raised—of becoming irritated or angry. Picture a fighting cock with his cape feathers bristling. And we've already noted that the word for cock in almost any language can refer to the penis. And we catch chickenpox. Why put the chicken into the name of a disease? Because it's weaker, less brave than smallpox.

The roosters depicted in cartoons are a varied bunch. Two of the more notable are Super Chicken, who starred in several *George of the Jungle* episodes along with his lion sidekick, and Foghorn J. Leghorn, featured in Disney's Looney Tunes and Merrie Melodies. Super Chicken is a feisty little thing in a blue cape. But, doggone, when he sings his song, he clucks like a hen. Foghorn J. Leghorn is a great big, barrel-bellied white leghorn rooster with a distinctive cape of brown feathers that he might have stolen from a Rhode Island Red. He starred from the mid-1940s to the early '60s and was resurrected for a few cartoons in the 1990s. The classic folktale "The Little Red Hen" was picked up by Disney in the 1930s and given the title "The Wise Little Hen." This cartoon marks the first appearance of Donald Duck, who was accompanied by Peter Pig; they were the two animals who tried to evade work when the

little hen asked them to help. We can't overlook Camilla, the Muppet hen with blue eyelids. Nor can we ignore the goofy animated 2000 film *Chicken Run*, in which head hen Ginger manages, with the help of a dashing rooster, to save the chicken house on Mrs. Tweedy's farm from being turned into a pot-pie factory.

Chickens also star in the 2001 PBS film *The Natural History of the Chicken*. But expect only

Cartoon rooster »

a tad of natural history in regard to the bird. Rather, the film documents the sometimes sweet, sometimes sentimental, sometimes completely wacky responses of human beings to the birds that they do—and in one case, do not—fancy. After crediting Ulisse Aldrovandi for being a premier proponent of chickenkind, the film begins by showing an awesome act of CPR. In the winter, Maine in the throes of a nor'easter can be unkind to chickens. One of a flock, the hen Valerie, is found outside frozen stiff under the porch. Her person, Janet Bonney, brings her inside and discovers that there is a faint pulse in her neck. The next step, mouth to beak resuscitation, brings her back to life. After a week of recuperation in a baby's crib, Valerie is up and about, good as new. Another chickener does arm-flapping, foot-lifting, crowing-rooster dances to show how a cock advertises his services—food-finding and mating—to his hens. A chickenette (that's all that I can think of to call her) keeps a pet white Silkie rooster, Cotton, whom she bathes and blow-dries, whom she carries with her everywhere in a basket, whom she diapers, and for whom she has made a Silkie-sized car seat. She calls him "my soul mate." Several cautionary scenes are interpolated amid the varied views of human nature. One shows a super-large hatchery for broiler chicks—a CAFO—along with the vast and vastly crowded barn in which they grow to market weight. We also see a hen house—another CAFO—that holds more than two hundred thousand laying birds. In contrast, we are introduced to a farmer who pastures his hens, which are cooped at night but range free by day. His hens lay hither and yon; gathering the eggs is the job of two little girls, who must feel as if they're on a daily Easter egg hunt. When the hens become too old to lay, they are slaughtered in the yard, dressed out, and roasted.

The people who are chicken antagonists are the people who live in an Ohio neighborhood beset by non-stop crowing for a good fourteen hours, starting at five in the morning. One resident, newly arrived, raises fighting cocks. (It's evidently okay to raise fighting cocks in Ohio, though it's not legal there nor in the forty-nine other states to let them actually engage in combat.) Each bird is tethered to a metal hutch that looks like a large mailbox, and he'll clamber atop his hutch to crow defiantly at the competition, which amounts to more than a hundred blustering birds making more than twenty-thousand crows each and every day. "It was like Chinese water torture," says a besieged neighbor, "a thousand little drips just falling on your forehead." The neighbors finally win a lawsuit citing the breeder as the creator of a

« Modern Game rooster

nuisance. He is required to reduce his flock to only five birds.

Only the very last segment of the film engages the viewer in the chicken's natural history when it zeroes in on the hen's overwhelming protectiveness of her chicks. A hen with chicks is the avian equivalent of a mother tiger. We meet Liza, a white Silkie bantam picked on by the bigger hens, who is given her own bantam-sized chicken house, where, indeed, she lays eggs and plucks out her breast feathers so that the warmth of her body will aid incubation. Six chicks peck their way out of her tiny eggs—two yellow, two mottled gray-brown, and two black. When a hawk comes scouting the chicken yard, the hens hide, but Liza remembers her chicks vulnerable out there in the grass. She runs to them and summons them under her wings. The hawk strikes. The chicks are saved because of their mother's instinctive response to their danger. As it happens, the hawk misjudges her size because of her fluffiness. Silkie seven, hawk zero. Only a few white feathers float on the grass.

As models of domesticity, hens dwell busily in nursery rhymes:

> I had a little hen,
> The prettiest ever seen;
> She washed up the dishes,
> And kept the house clean.

She went to the mill
To fetch me some flour,
And always got home
In less than an hour.
She baked me my bread,
She brewed me my ale,
She sat by the fire
And told a fine tale.

This is hardly the first time that the word "hen" has been used to mean "woman." Hen parties have been held since human time began. In the next nursery rhyme, we meet a hen, who is a hen and like all hens can't count:

Chook, chook, chook, chook, chook,
Good morning, Mrs. Hen.
How many chickens have you got?
Madam, I've got ten.
Four of them are yellow,
Four of them are brown,
And two of them are speckled red,
The nicest in the town.
Chook, chook, chook, chook, chook,
Cock-a-doodle-doo.

"Chook"—rhymes with "book"—is a colloquial word for "chicken" in Australia and New Zealand. I wonder if it had not been taken to those countries by English people familiar with the little rhyme. And here's another domestic hen, overdoing what her kind does:

Hickety, pickety, my black hen,
She lays eggs for gentlemen;

Gentlemen come every day
To see what my black hen doth lay.
Sometimes nine and sometimes ten,
Hickety, pickety, my black hen.

Nor are cocks left out. Here's one performing his time-honored task:

The cock crows in the morn
To tell us to rise,
And he that lies late
Will never be wise.

≫ Australorp hen

Eggs, too, have their place in nursery rhymes, to wit the tragedy of Humpty Dumpty.

And they show up in song: "I gave my love a chicken without a bone." The song holds the answer: "The chicken in the eggshell, it has no bone." Another well-known song is "I Love My Rooster," popularized in a children's album by cowboy singer Tex Ritter.

The tune and lyrics come from a folksong far older than Ritter.

I love my little rooster, my little rooster loves me,
I'll cherish my little rooster by the green bay tree.
My little rooster goes cock-a-doodle-doo
Doodle-doo, doodle-doo

« Humpty Dumpty

The hen figures in the next verse, then a duck, a guinea hen, and a pig. The song was included in the John Quincy Wolf Folklore Collection at Lyon College, Batesville, Arkansas.

Chickens cluck and crow in a coopful of stories. They figure in classic tales like "The Little Red Hen." No one knows quite where she first asked the other animals in the barnyard for help. I've seen a version that uses the word "corn" for "wheat" and attributes it to British tradition, but there's a good probability that Russia was the country in which she planted the grain, made the flour, baked the bread, and ate it all by herself because the duck, the pig, and the cat had refused to help. Russia is also one of the native lands

⊼ The little red hen

of Baba Yaga the witch, who lives in a cabin atop dancing chicken legs. Another beloved children's story, "Chicken Little," does boast an author, Merri Beth Stephens. A slew of other people have claimed it, but their contributions have been not the story itself but rather illustrations. The characters are irresistible. Bopped by an acorn, Chicken Little (called Chicken Licken in some of the versions appropriated by artists) thinks the sky is falling and rallies her family and a friend—Henny Penny, Cocky Locky, and Turkey Lurkey—to tell the king of the imminent disaster. En route, they encounter Foxy Loxy, who sees in them a delicious opportunity for dinner. But the king's hunting dogs, barking and baying up a storm, arrive in the nick of time. From then on, Chicken Little carries an umbrella when she walks in the woods, and if an acorn were to fall, it certainly wouldn't bop her.

One of my favorite chickens crows in Chaucer's *Canterbury Tales*. "The Nun's Priest's Tale" tells of a flock kept by a widow, who had two little girls. The birds lived in a yard on her small farm along with three cows, three sows, and a lone sheep.

> And in the yard a cock called Chauntecleer,
> In all the land, for crowing, he'd no peer.
> His voice was merrier than the organ gay

On Mass days, which in church begins to play;
More regular was his crowing in his lodge
Than is a clock or abbey horologe.

He was a most amazing bird to look at—coral-red comb, black bill, blue legs and toes, white spurs, and burnished gold feathers. His paramours numbered seven, and of them, his favorite was Mistress Pertelote.

But nightmares auguring no good assailed him. Pertelote prescribed a regimen of laxative herbs—laurel, hellebore, spurge, dogwood berries, and more—to purge his body and his brain of bad dreams. But Chauntecleer told her of sleeping men whose dreams presaged murder or drowning, King Croesus dreaming of his hanging, Hector's wife dreaming that her husband's life would be lost in the Trojan War. And he stated flatly that laxatives are poisonous. He'd rather be riding her, although he can't do that at night because their perch is so narrow. The next morning he trod her twenty times before 9:00 AM.

Come spring, Chauntecleer and his wives walked merrily in their yard. But his ominous dream was about to come true. After waiting patiently for several years, a col-fox—a fox with black-tipped tail, feet, and ears—had breached the hedge and entered the yard. The animal hid in the bushes waiting for the chance to pounce on a rooster dinner.

At this point, the nun's priest comes in with an aside to the effect that the cock should not have listened to female counsel that his dreams had no significance. After all, look what happened when Adam listened to Eve.

A modern Chauntecleer and his paramours. »

Carefree, Chauntecleer and his wives bathed in the chicken yard's dust.

> And so befell that, as he cast his eye
> Among the herbs and on a butterfly,
> He saw this fox that lay there, crouching low.
> Nothing of urge was in him, then, to crow,
> But he cried "Cock-cock-cock" and did so start
> As man who has a sudden fear at heart.

Seeing that Chauntecleer was about to run, the fox spoke up, swearing that he was the cock's friend and that he'd come to the yard only to hear him sing. His father had sung magnificently. Could Chauntecleer sing so well? And the cock gave it a try:

> This Chauntecleer stood high upon his toes,
> Stretching his neck, and both his eyes did close,
> And so did crow right loudly, for the nonce;
> And Russell Fox, he started up at once,
> And by the gorget grabbed our Chauntecleer,
> Flung him on back, and toward the wood did steer.

The hens raised a ruckus. Pertelote cried loudest of all. The noise summoned the widow and her daughters. It rallied all the dogs in the neighborhood. Everyone, even the cow, her calf, and the hogs, ran toward the wood. All of them cried, shouted, barked, mooed, and snorted. The fox had by then reached the edge of the dark woods.

The crafty fox »

Chauntecleer told him that if he were the fox, he'd tell the pursuing mob, "In spite of you this cock shall here abide. I'll eat him, by my faith."

The fox replied that it would be done. A great mistake to open his mouth, for the cock broke free and flew up into the crown of a tall tree. The fox called out an apology for frightening the bird but that it had been done with no malice aforethought. But the cock had learned his lesson. He told the fox that he'll never again be flattered into singing and closing his eyes. The moral of the tale, of course, is don't succumb to flattery.

"The Nun's Priest's Tale" closes with a delightfully bawdy epilogue. The host of the pilgrims traveling to Canterbury says to the nun's priest:

> But, truth, if you were secular, I swear
> You would have been a hen-hopper, all right!
> For if you had the heart, as you have might,
> You'd need some hens, I think it will be seen,
> And many more than seven times seventeen.

In Shakespeare's *Hamlet*, a cock is magical. He performs his traditional and instinctive task by heralding daybreak and also dispels a ghost. In the night, Marcellus and Bernardo, two soldiers on night watch at the court of Denmark, have witnessed the apparition of Hamlet's father's ghost. They summon Hamlet's longtime friend Horatio to see this uncanny phenomenon. The murdered king does indeed appear in the darkness but vanishes. Bernardo says that the specter was about to speak, but the cock crew.

The nun's priest

HORATIO
And then it started like a guilty thing
Upon a fearful summons. I have heard,
The cock that is the trumpet to the morn
Doth with his lofty and shrill-sounding throat

Awake the god of day; and, at his warning,
Whether in sea or fire, in earth or air,
The extravagant and erring spirit hies
To his confine; and of the truth herein
The present object made probation.

MARCELLUS
It faded on the crowing of the cock.
Some say that ever 'gainst that season comes
Wherein our Saviour's birth is celebrated,
The bird of dawning singeth all night long:
And then, they say, no spirit dares stir abroad;
The nights are wholesome; then no planets strike,
No fairy takes, nor witch hath power to charm,
So hallow'd and so gracious is the time.

Horatio replies that, though he's heard of this cock-engendered armistice between the sacred and the profane, he admits to some skepticism. All exeunt to fetch Hamlet. Perhaps the ghost will speak to his son. The next night, he does tell Hamlet that he was murdered. One death leads to another and more. The cock, nonetheless, has done his own task and done it well.

Chickens are likely to inspire silly verses, in which, for example, a writer dreams of chickens exploring his body, nesting in odd places, and infesting his room only to wake and find real eggs atop his stomach. Chickens deserve better, and they've received it.

In "A Blue Ribbon at Amesbury," Robert Frost fondly lauds a prize-winning hen. The poem mentions Franklane Sewell (1866–1945), a notable painter of poultry, whose work is still used in the APA's *The American Standard of Perfection.*

Such a fine pullet ought to go
All coiffured to a winter show,

⌃ Sewell's painting of silver-penciled Plymouth Rocks

And be exhibited, and win.
The answer is this one has been—

And come with all her honors home.
Her golden leg, her coral comb,
Her fluff of plumage, white as chalk,
Her style, were all the fancy's talk.

It seems as if you must have heard.
She scored an almost perfect bird.

« White hen

In her we make ourselves acquainted
With one a Sewell might have painted.

Here common with the flock again,
At home in her abiding pen,
She lingers feeding at the trough,
The last to let night drive her off.

The one who gave her ankle-band,
Her keeper, empty pail in hand,
He lingers too, averse to slight
His chores for all the wintry night.

He leans against the dusty wall,
Immured almost beyond recall,

A depth past many swinging doors
And many litter-muffled floors.

He meditates the breeder's art.
He has half a mind to start,
With her for Mother Eve, a race
That shall all living things displace.

'Tis ritual with her to lay
The full six days, then rest a day;
At which rate barring broodiness
She may well score an egg-success.

The gatherer can always tell
Her well-turned egg's brown sturdy shell
As safe a vehicle of seed
As is vouchsafed to feathered breed.

⩖ Rhode Island White,
a hen that lays a brown egg.

No human specter at the feast
Can scant or hurry her the least.
She takes her time to take her fill.
She whets a sleepy sated bill.

She gropes across the pen alone
To peck herself a precious stone.
She waters at the patent fount
And so to roost, the last to mount.

The roost is her extent of flight.
Yet once she rises to the height,

She shoulders with a wing so strong
She makes the whole flock move along.

The night is setting in to blow.
It scours the windowpane with snow,
But barely gets from them or her
For comment a complacent chirr.

The lowly pen is yet a hold
Against the dark and wind and cold
To give a prospect to a plan
And warrant prudence in a man.

She is such a fine hen that she could well be the progenitor of a magnificent new breed. The young keeper, however, sees in his blue-ribbon hen a caution that prudence befits human-kind. He looks to the ancient ideal of balance, of moderation in all things.

The epitome of the poem inhabited by chickens is surely that written in 1923 by W. C. Williams:

so much depends

upon

a red wheel

barrow

glazed with rain

water

beside the white

chickens

The poem looks as if it's a simple sentence. But it's filled with color and light-reflecting wetness. Imagine what the poem might have been like if its last word had been "lilies" or "downspout." "Chickens" was the inevitable choice, investing these sixteen words with breath and a steady heartbeat.

CHAPTER FOURTEEN

Chicken People

Cities, suburbs, the countryside, chicken people are to be found everywhere. They are happy and obsessed. Come, meet some of them.

Grandma Moses

As a bride, folk-artist-to-be Anna Mary Robertson Moses (1860–1961) came to the Shenandoah Valley to live in a four-square, two-story brick house in Verona, Virginia. The first fall that she and her husband lived there on their farm of a hundred or so acres, she bought twelve old hens from a neighbor for six dollars. Two of those old hens became broody and started to set. In her book *Grandma Moses: My Life's History*, she wrote that she sent three dozen of her eggs to a neighbor "and exchanged them for brown leghorn eggs to put under my setting hens. They did fine. I continued all the rest of the spring until I had 118 little chicks from my 12 old hens. They are lots of care but good company." She employed a novel means of keeping her old hens from going on escape. She'd tie a hen to a

Grandma Moses »

brick, then put her and her chicks out in the sun. If it rained, she'd pick up hen and brick and take them to the woodshed. Right smartly, the chicks would follow.

After she began painting—she was in her seventies at that point—she remarked that if she hadn't become an artist, she could always have made a living raising chickens.

White Hen, Rescued

Downtown in a local picture-frame shop cum camera museum, I visit with Sherrie Breeden, wife of the proprietor, and she shared this story: Once upon a time in the early 1960s, her grandparents, Herbert and Cary Jean "Jim" Collins, came across a wrecked chicken-transport truck as they were driving back to their home in Charlottesville. Dead chickens, injured chickens, and chicken cages were strewn across the road and its margins. One white hen was walking and pecking through the wreckage. What to do but pick her up and take her to their yard on a busy street in Charlottesville? She spent her days tied by a string to a forsythia bush in the backyard and her nights roosting on a pipe in the cellar. Eventually, a good five or six years after her rescue, she died peacefully of old age.

Charlottesville, not incidentally, has never had an anti-chicken ordinance, though it has gone so far as declaring roosters out of bounds. The city now has an organized group of chickeners, who call their organization CLUCK—the Charlottesville League of Urban Chicken-Keepers. I'll bet, though, that their hens are not tied to forsythia bushes nor do they roost in anyone's basement.

Cleveland Farm Animals

Doug Johnson, a jovial Clevelander in his early sixties, is an urban chicken-keeper. As he tells his story, he smiles broadly and hugs one of his girls, a Rhode Island Red, of which he has three, along with a Golden Buff and a Silver Laced Wyandotte.

Doug Johnson and one of his girls »

"This summer, my wife caved in and let me get chickens. Cleveland changed the zoning rules a year ago, so I am now doing 'Chickens in the Backyard' for real. It's a no rooster policy—that's the *law*—so it's girls only, but I do expect eggs in a month or so. The neighbors have found it interesting and, yes, they are all okay with it.

"Although I knew the law had changed, I didn't know there was a permit needed.

"So, one day there was a card from the health department. Turns out I just needed to fill in a bunch of paperwork and pay a fee—natch. So I am now the proud bearer of Cleveland Farm Animal License number 17. Apparently, it's not a big hit—or people aren't filing the forms."

Sky Burial

Lilly Golden keeps a flock of free-range chickens on a small farm at the edge of the Catskills with her husband and two daughters. It's February, and she writes, "Our chickens are laying now that the days are lengthening. We have nine in total now, including an enormous accidental rooster named Sally, an Araucana. Five hens are newish, a couple of years old. They lay brown eggs. Our middle-aged group is about three or four years old. They lay bluish-green eggs. Our eldest, almost seven years old, the last of the original group, is Cheeky, an affectionate auburn and gold Araucana with big cheek feathers. She lays lovely olive-green eggs. It's a rare event at her advanced age, but all the more exciting when it happens."

Then she tells me a rooster story: "Several years ago, before Sally's time, we were given a surprise addition to our coop while we were out of town—a bantam rooster. He was a nasty little guy once he came into his own. He attacked me mostly, unless I had a dog or horse at my side. When he turned on the girls one day, that was it. We caught him after a Great Rooster Chase only after he became

Chickens improving Lilly's landscape »

Rose and Cheeky »

cornered in a barn. He cried piteously when he was captured. I felt sorry for him, but we gave him back to the people who gave him to us, and they 'set him loose' in the woods. No doubt, a fox or coyote or mink had him for lunch."

Chicken-keeping has its tragedies. "We recently lost Midnight, who was also of the first batch," says Lilly. "She was a mostly black Araucana with gray flecks. Our ground is too frozen to dig a hole for a proper burial. So I snowshoed out into the woods quite a ways and put her up in the wide arms of an old tree, where another chicken had been laid to rest a few winters ago."

Sky burial is a ritual in Tibet and has been so for many Native American people. Exposing the body to the elements and other creatures is a splendid way of recycling life. Midnight, RIP.

Oberon, Winnifred, Eleanor, and Charlotte

Tristan Walker is a tall, serious young man in his mid-twenties. He lives in St. Louis in the densely populated Hi-Pointe neighborhood of the larger Dogtown neighborhood that is roughly bounded by Interstates 64 and 44. Hi-Pointe is located atop one of the highest spots in the city and affords Tristan "an awesome view" of the city of St. Louis spread out on its alluvial flood plain. A self-styled urban farmer, he's castled in a shotgun duplex of about six hundred square feet. The lot is fortunately large for the house and the neighborhood. He keeps a good-sized garden, mainly vegetables with some flowers, between his house and the fenced portion of the yard. His chickens live behind the fence. They came with the house. St. Louis allows four pet chickens—hens only, not roosters. The city does not require a permit.

"When I got my house," Tristan says, "I was happy to take over chicken care." Along with his house, he acquired two Ameraucana hens, gray Winnifred and wheaten Eleanor; one matronly black Australorp hen, Charlotte; and—St. Louis be damned—Oberon, an Amer-

« Eleanor, Winnifred, Charlotte, and Oberon

aucana rooster. The bird has never heard of the king of fairies but, doggone, he knows that he's king of the hens.

"But you're not allowed a rooster!"

"Actually, it's all right." he tells me. "A building inspector came when I bought the place. She asked if any neighbors had objected. No. So she gave me a pass. But she did cite me for a broken window."

He has a coop built on the end of his garage. "There's an entirely caged-in run they may access at the rear of the garage. I let them free-range in the whole backyard during the day. Seeing as how the walls for half of my garage have been removed, they have a protected porch. They spend their days here during the winter. They are definitely the most spoiled chickens I've ever heard of.

"The most striking characteristic about chickens," Tristan has noticed, "is their sexual dimorphism. Roosters are almost a parody of masculinity, while hens are very feminine. Oberon's mode of interaction with almost anything that enters the backyard is one of aggression and challenge. If he recog- nizes a hawk or another obvi- ously dangerous predator, he will hide. But people, dogs, cats, and some birds get challenged. If you back down, a chase will ensue, and you will get spurred. He defi- nitely has the well-being of the

Oberon »

hens at the top of his priority list. He always announces to them when he finds tasty morsels in the yard." He also spreads his wings at dusk and flaps his hens into the safety of their coop.

"I find it fascinating to try to figure out what parts of their behavior are instinctual and how much is based on their limited reasoning capacity. They are all naturally frightened of raptors, but Oberon has learned that bare skin is more sensitive than shoes or clothing. So, he'll seek out tender spots to peck if he is displeased. The hens have a pecking order, of course, and enforcing that seems to occupy much of their time. When mating season starts, they will distract themselves from the constant hunt for food to seduce the rooster by shaking their bustle, but that's their only behavior that's not directly related to eating, sleeping, or maintaining the status quo.

"One of the hens, Eleanor, is unique. She was attacked by a raccoon when she was young. So, her previous owner, a high school biology teacher, took her to work every day in a basket. As a result, she's very accustomed to humans. She has been known to hop into laps looking for treats, and she knows that nestling her head under your arm elicits happiness from people.

"And let me tell you how Ellie vanquished the possum. Four AM, there's an awful ruckus in the chicken house. So I get out of bed—oh, it's cold. Oberon and the girls are hysterical. And there was the repeat offender—the possum. Not in the coop but trying."

Just a month earlier, Tristan had drop-kicked the possum out of the yard. This time, the villain was given the death penalty. Tristan went into the coop and tried to calm his birds.

⤳ Possum

"I sat on the swing rocking back and forth and cuddling Ellie. She decided to climb on my shoulder. Then Oberon decided, 'Me, too,' and climbed on my other shoulder. I'm wearing chickens like epaulets. Ellie got down pretty soon and started goose-stepping around the coop—Ellie triumphant. It was as if she had routed the possum all by herself.

"Once Oberon got into a fight with a dog. A dachshund squeezed through my fence, and Oberon ran up and attacked it. Given that dachshunds are bred for hunting things that run along the ground, it was not going well for the rooster. He lost his tail plumes and a hefty chunk of flesh out of his back. My housemate and I made it out to the backyard in time and violently ejected the dog from the yard. We irrigated the wound and spread some antiseptic on it. It healed up over the course of a month. The reptilian nature of birds, chickens in particular, seems to grant them impressive regenerative powers."

"Tristan, would you, could you eat your hens when they get too old to lay?" I asked.

"All my hens are coming up on four or five years, and their productivity is starting to wane. I'll let these hens depart in a natural manner, for they aren't really mine. They just came with the house. Any other chickens I hatch may very well get turned into soup once they stop laying or start fighting other roosters."

Winnifred, Eleanor, Charlotte, and the uxorious Oberon—Tristan is happy with his birds. Nor are these the only winged creatures that he will keep. He says, "I'll be starting a beehive this spring."

One Smart Chicken

"Lots of non-chicken people talk about how dumb chickens are," says Ellen Von Seldeneck, known as E. V., who runs Mantis Gardens, a landscaping business in Asheville, North Carolina. "But something that happened not long ago makes me question that for sure."

It was in 2009 that Asheville changed its ordinances to allow chicken-keeping. Restrictions are in place: A permit is necessary, chickens cannot be located closer than one hundred feet to neighboring households, chickens must be confined in a pen that the city will inspect, and the poop must be scooped and put into a sealed container. It's still illegal to keep a rooster, but enforcement of the ordinance is complaint-driven. In what E. V. describes as a very urban neighborhood, she and six other households have established a community garden in a field at the end of the street and a chicken cooperative in the lot right next to the field. A big coop, a small protected run, and a larger run with more freedom have been installed for the birds. Each household has its designated day of the week for chicken care.

"What really excites my passions," E. V. says, "is growing food, whether it be from the ground, the chickens, or the bees." (Yes, like Tristan, she cannot resist wings.) "I especially love doing this in an urban setting. The way I see it is, our city, like most, is full of small pieces of very farmable land that is just sitting useless, usually in grass. If everybody could grow just a little of their own food, what a difference it would make! Most of my neighbors don't go to farmers' markets. The food costs three to four times more than the cheap, low quality, non-organic they drive to get. There's just so much wrong on so many levels."

I ask her if she'd eat her chickens.

"I'd thought that I'd eat them easily once they're done laying, but now I'm not so sure. It's a bit too personal. But I definitely want somebody to eat them. So, when they are really done, our neighbor, Norman, will most likely kill and eat them for us."

Then she launches into her grand chicken story. "Our chickens live at the end of our dead-end street, several houses away from me. They have lots of moms, but we are their number one caretakers. Even when they manage to break out of their fencing, they stay basically right there in my neighbors' side-yard. They can't see the rest of the neighborhood. That's their world—this little lot.

"One day I was at my house about dusk, and I hear a strange disruption at the door. When I open the door, one of the Rhode Island Reds is standing right on my porch. Well, I knew immediately that something was wrong. It's so weird that she came that far away from the coop. I shut the door so she can't walk in the house, and I proceed to change shoes and put on a jacket. I hear another clamor at the door and look through the peephole. She's throwing herself at the door! She's flying into it, stumbling back, and throwing herself again to get my attention, which she already had.

"I pick her up and run down to the coop afraid to find a dead chicken and a very happy dog. But what I find is that the door to the coop had blown shut during the day. Now it's dusk, and the chickens are locked out, desperately trying to figure out a way to get into the coop—all thirteen of them just freaking out. I guess one smart chicken decided to walk down the street and knock on my door so I'd come fix it."

Admirable! To E. V.'s way of thinking, there's no such thing as a dumb cluck.

Six Chicken Lovers

AnneMarie Cumiskey, her husband, Michael, and their four children live in the country near Scottsville, Virginia, a town located on a miles-long horseshoe bend of the James River south of Char-

⌃ Alpha Girl, the hen that knocked on the door

lottesville. When I began thinking serious chicken thoughts, I found her on Craigslist as a purveyor of a sturdy, pocketbook-friendly coop made of Douglas fir and shingled with asphalt. Suitable for five or six hens, it seemed just right for my backyard.

The Cumiskeys style themselves Six Chicken Lovers, a name coined by Michael, who works as an information technologist. AnneMarie deems him Chicken Lover #1. Their children are Chicken Lovers #2 through #5. Kathy, whom AnneMarie calls "my chicken girl," is the eldest; she takes care of the birds every day. Kathy also sells the eggs and contributes part

of her earnings toward the upkeep of the birds. Emily, "my animal and nature lover," comes next, followed by Paul, just old enough for elementary school, and Angela, not long out of the toddler stage. "I'm #6," AnneMarie says, "aka Mom." And Mom is home-schooling her offspring. She'll soon be introducing

Angela cuddling a bantam »

them to Latin. What better place to start than with *gallus* and, for a hen, *gallina*?

The Cumiskeys started keeping chickens when they moved into the country a decade ago. "Michael wanted chickens," she tells me. "He thought it would be a good hobby for our then small family—we only had one child at the time. He did the research and bought five Rhode Island Reds. Recently, he told me that he had never actually seen live chickens before. He grew up in the suburbs of Northern Virginia. I grew up in Trinidad, in the Caribbean, and my aunt used to raise chickens. So, I was a *bit* more familiar with them."

At the time that I write, the Chicken Lovers have twenty-seven birds. The flock abounds with color and variety: black-and-white barred

≈ Emily holding a Speckled Sussex

Dominiques, an American breed that's daintier than the similar Barred Rocks; Australorps with shiny black feathers that hold a touch of green and purple iridescence; Buff Orpingtons, the name of which denotes their creamy coloring; Speckled Sussex, an uncommon breed with white-spotted black feathers and an origin in England; Marans, a French breed with cuckoo coloring, which has nothing to do with cuckoo birds but rather denotes a black-and-white barred feather pattern; hybrid Red Stars that lay huge brown eggs; and Ameraucanas with the gene for laying blue eggs. As if these were not enough, the Chicken Lovers also keep ginger-red Old English Game bantams in a small coop elevated on pilings and blessed with an attached run.

Roosters? Oh, yes, six of them. "No one seems to want roosters, so we have a fair number of them," AnneMarie tells me. "We don't mind the crowing. After a while the sound fades into the background, and we don't even think about it—unless I'm outside on the telephone." But she does find one aspect of roosterdom that worries her: aggressiveness. "A few years ago, a rooster attacked Kathy's ankle with his spurs, and I had to take her to the emergency room. That was the rooster we killed."

Most of the birds that are getting on in years have been named—Sarah, Whitebee, Shallow, M, and Iris. A family joke has it that the Cumiskeys are running an old-age home for chickens. A few of the younger birds have also acquired names—Star, Charlie, Barbara—and the Australorp pair is Mr. and Mrs. Tuft. Except for the attack-rooster, they do not kill the members of their flock. "Aside from which," AnneMarie says, "chicken is relatively inexpensive in the supermarket."

The live birds give them great pleasure. They roam the yard but are given supplements—laying pellets and kitchen scraps. They help themselves to rooting through the compost pile. "My husband enjoys watching them free-range. He finds their antics very relaxing. The kids enjoy holding them and playing with them. Kathy likes to walk about with a well-trained bantam hen perched on her arm. As for me, I'm most delighted with the eggs and fertilizer."

The chickens are the clucking, crowing heart of a larger farming enterprise. The farm sports a kitchen garden, of course, where the family grows a full range of vegetables, from tomatoes through squash and beans. An acre has been planted with edamame, and a thousand blackberry, blueberry, and raspberry bushes occupy a field. AnneMarie's coop sales help to support the berry-bush enterprise. The crop of edamame will sell annually at the local farmers' market in the fall. Along with the produce, the farm supports animals other than the chickens. Two dwarf Nigerian goats, Poe and Honeycomb, make their home here, along with two lambs, Snow and Oreo. Nor are these all. The Cumiskeys' livestock includes six heirloom Bourbon Red turkeys and twenty bobwhite quail, all raised from day-old poults. Forty weeder-goose goslings—White Chinese and Africans, the latter with bold black knobs just above their beaks—have joined the rest. Weeder geese? Yes, indeed. Young geese turned

Dwarf Nigerian goat »

loose amid crops eat the weeds that suck up nutrients and leave the vegetables, flowers, and berry bushes alone. And they work from dawn till dusk and even later if the moon is bright. The Cumiskeys will put them to work not only in the acre of edamame but also amid the berry bushes. Because they tend to goof off in the weeding business as they age, they'll be sold for Thanksgiving and Christmas feasts, come fall.

⌃ African weeder geese

Chickens Galore

"My chickens are my love," says Brenda Smith, a round and ebullient great-grandmother wearing jeans and crowned with red curls. She's in her early sixties. As it happens, she has a lot more than chickens to love, for she's the doyenne of a forty-eight-acre farm that she's named Bonnie Grace. Situated out in the county not far from my town of Staunton, it offers fine views of the Blue Ridge Mountains. Christian's Creek, named after a pioneer family, flows through her land bringing it sweet water. Her old blue farmhouse, situated a good quarter-mile from the paved road, is surrounded by animals—cows in several pastures, ten outdoor cats, three lambs *baaing* loudly and chasing one another around an outbuilding, and chickens galore. It's not just the chickens that she loves but all the other meowing, bleating, and mooing mob of creatures with which she has surrounded herself.

On a day early in March, I arrive at Bonnie Grace with the friend who suggested that I get in touch with Brenda. Our greeting is a full-throated crow from Jack Be, a snowy white Leghorn rooster with a plumy tail. He rounds up his hens by running toward them and flapping his wings to assemble them where he wants them. Just as he's supposed to do, he keeps the chicken yard in the classically grand order described by chickeners from Columella to the Reverend Mr. Dixon. Brenda has described Jack Be as "a bully, but he's *cute*." One of the hens in his harem is named Nimble; another is Quick. Candlestick, alas, died not long ago.

"Do they all have names?" I ask.

She waves at two red hens. "One's Henny Penny, and one's Red, but I can't tell them apart. Look over at the barn—one hen over there. She's a loner, won't join the flock. Now, come meet Christmas. He was just a little guy when I got him on Christmas Day. Now he's a bit more than two months old. Have to keep him cooped up so Jack Be doesn't get him."

Christmas, a Leghorn, lives in the shed that also houses the lambs, one of whom rises from a big feed tub when Brenda opens the door. Nor is Christmas the only chicken resident in the shed. Three small, mop-headed Polish chickens live there, too. Brenda couldn't resist them when she saw them for sale at a farm exposition. Two, Rufus and Rastus, are roosters; Ruthie is a hen. All are buff-laced in color—light gold feathers with white speckles. Their jaunty crest feathers bounce up and down as they scurry around the shed.

"Those roosters think they're sheep," Brenda says, arms akimbo, hands on her hips. "I had to give them a talking to. 'You're not sheep,' I said. 'You're roosters. You're supposed to roost.' But they didn't, and I had to give them another talking to. They still didn't listen, so I just plain put them up on the roost. Next thing you know, they're all down, running around like the lambs—except for one who's still up there. Had to give them another talking to."

The lambs, but not the roosters, scooted out the moment that the door was opened. I've never seen such coloration in lambs. Leroy and Lizzie are two-week-old twins, charcoal in

⌃ Brenda and Polish chickens

color with white faces; Ben, four weeks old, is charcoal frosted with white all over his flanks. "You'll have to spin their wool after they're shorn," I say.

"They never grow wool," Brenda replies. "They're hair sheep bred for meat, not wool."

They were given to Brenda, who is known far and wide as a soft-hearted sucker for animals. She's a one-woman rescue operation for anyone who wants to be disencumbered of unwanted animals. She points to a group of cattle grazing on a hill to the south. "Up there's Charlie Bull. He ain't a bull, though, he's a steer. When he was given to me, he wouldn't eat. He just laid out in a building with his tongue hanging out. My handyman tube-fed him for six weeks. One day, he went up there expecting to see a dead steer. No! Charlie Bull was up and running around." One of her dairy cows, Big Blue, fed five calves last year. Charlie Bull, the

« Hair sheep

calves, the hair sheep, Jack Be, and most of her hens were discards.

"My favorite hen is Izzy. Let me get her."

I'll learn later that Izzy came to her as a small pullet not more than a month or so old. She'd been given to a woman who had no coop outside and no inclination to let a chicken wander around inside. Big-hearted Brenda provided an ideal solution and almost certainly saved Izzy's life.

Brenda is back in a trice bearing a huge white hen. "She's a broiler hen—big breasted. So big she can't walk." Izzy should have been someone's dinner, but for reasons unknown, she was spared slaughter at the age of six weeks. I think of my daughter's immense rooster, who could walk, a feat that spared him Izzy's indignity. Food and water close at hand, she sits inescapably in her own poop. It clings to her tail feathers in big clumps the size of shooter marbles. Brenda nuzzles the bird, which closes her eyes in something that looks like chicken bliss. She lives, and that's what counts.

Brenda's chickens are well-socialized. Her cats wind around our legs and purr. The lambs show no fear, not even of the FedEx truck that has come to deliver a large, clanking package. She picks up Leroy Lamb and cuddles him till the truck is safely back on the road. She radiates love.

Addendum: Christmas

Mid-May: it's time to decoop Christmas, to let him out of the shed that he shares with the lambs and the Polish chickens, for he's reached a good size, enough to hold his own in the presence of the harem-master, Jack Be. Holding his own may, of course, consist of running away, but having practiced going in circles with the lambs, he's good at running. When Brenda opens the shed door, Christmas pokes out his head and looks uncertainly about. Brenda beckons, and he decides at last that he'll leave the roost.

Come evening, Jack Be and his paramours have cooped themselves. Christmas is due back at the shed, but Brenda can't find him anywhere. She calls and calls. No answer. He's in danger from night predators if he's not sequestered. Finally, she hears a soft *puck-puck* in a corner of the big barn. He's there beside the hen that refuses to join the flock.

"Christmas. Christmas. Come, Christmas." She tries chicken-talk: "*Puck-puck-puck.*"

He refuses to budge from his chosen spot. *Puck-puck, puck-puck*, he says. He's not answering Brenda. He's talking to the hen.

After long minutes of standoff, Brenda persuades him that it's time to go home.

He continues his conversation with the hen. It's as if he's apologizing to her for leaving. He's surely telling her that he'll be back. And he follows Brenda to the shed.

"Well!" Brenda says. "Christmas is in love."

After this first foray into the wide world, decooping becomes a matter of course for Christmas. As big as Jack Be, he's held his own in a few tussles. But he has other matters on his mind and invents a job for himself. Brenda has acquired a whole flock of hair lambs; if one or two should stray from the mob, Christmas runs at them flapping his wings as if he's rounding up hens. Nor does he cease until the lambs have rejoined the others. It may be that Christmas still thinks that he's a sheep. He's certainly a shepherd.

One noon, Christmas disappears. Brenda looks everywhere. No carcass. No feathers. Jack Be can be absolved of responsibility. Predator? Brenda finds consolation by telling herself that her lovely white rooster has moved himself to the chicken-yard of the farm next door.

A Glossary of Chickens

"I adore chickens," says Gary Whitehead, a dark-haired man with a closely trimmed dark beard and a slightly crooked mouth that looks as if it's trying not to burst into a smile. When he's not writing poems, painting with oils, or concocting crossword puzzles for the *New York Times*, he teaches at the high school in Tenafly, New Jersey.

"I raised backyard flocks of hens for a few years. I wish I could do it again, but it's just not possible. My yard's too small, the town doesn't allow it, and I have a dog with a known penchant for live chickens."

⌃ The sweetness of hens

But his imagination keeps chickens, and he has written a paean to them. He says, "One day, when I was at an artists' colony in the Adirondacks, a friend handed me a photocopy of a glossary from the back of a book on raising chickens. I wrote my poem after reading through it."

Here's "A Glossary of Chickens":

> There should be a word for the way
> they look with just one eye, neck bent,
> for beetle or worm or strewn grain.
> "*Gleaning*," maybe, between "*gizzard*"
> and "*Grit*." And for the way they run
> toward someone they trust, their skirts
> hiked, their plump bodies wobbling:

"*bobbling*," let's call it, inserted
after "*blowout*" and before "*bloom.*"
There should be terms, too, for things
they do not do—like urinate or chew—
but perhaps there already are.
I'd want a word for the way they drink,
head thrown back, throat wriggling,
like an old woman swallowing
a pill; a word beginning with "S,"
coming after "*sex feather*" and before "*shank.*"
And one for the sweetness of hens,
but not roosters. We think
that by naming we can understand,
as if the tongue were more than a muscle.

CHAPTER FIFTEEN

Chicken Cuisine

Time out of mind, chickens have been a global subject of culinary interest. The hatching ovens of Egypt and China bear witness to that.

Chicken

Early recipes, though, are hard to come by. To be sure, Pliny and others mention fattening chickens with the intent to eat them, but the early writers say little about just how they were cooked. It's not till the late fourth or early fifth century AD that classic recipes found a compendium. The collection is called *Apicius*, after Marcus Gavius Apicius, a Roman gourmet who lived in the first century AD.

Medieval cooks and those who benefited from their services used lots of chicken. To our great good fortune, some of their recipes, dating from the late fourteenth century, have been preserved. I'll spare you the Middle English of Richard II's cooks. The French *Le Viandier* or "The Food

Marcus Gavius Apicius »

Provider" by the pseudonymous Taillevent, is rich in chicken recipes, be the bird roasted, boiled, or simmered into soup. Luckily, translations exist. Here's one for Capon with Herbs. It will help to define the word "verjuice," sometimes spelled *verjus* (from the French *vertjus*—green juice), which denotes an acidic juice of grapes, crabapples, or other sour fruit. It's available today in red or white versions from some wineries. The red is bolder than the white.

Cook them in water, pork fat, parsley, sage, hyssop, costmary, wine, verjuice, saffron, and ginger as you wish.

This one gives directions for Chicken with Cumin:

Cook it in wine and water, quarter it, and fry it in lard. Temper your broth with a lot of wine, sieve, and boil. Add just a bit of ginger and cumin steeped in verjuice and wine. Take plenty of egg yolks, beat them well, and thread them into your pottage at the back of the fire. Make sure that it does not curdle.

We are fortunate to have a cookbook, *De arte coquinaria—The Art of Cooking*—that appeared in Italian about 1465 on the cusp of medieval-Renaissance cuisine. Its author: Maestro Martino di Rossi of Como, chef to the papal chamberlain and, later, in the Vatican. The book contains the first-ever mention of vermicelli as an ingredient. Its mentions of chicken and eggs are, of course, only the fifteenth-century latest in a long line of such recipes. In his initial instructions, Maestro Martino recommends roasting capons and pullets. He shortly thereafter presents a recipe for dressing a peacock with all its feathers so that, after it's cooked, it appears to be alive. To make it spew fire, place camphor-soaked cotton wool in its beak and light it just before serving the bird. At recipe's end, the Maestro suggests giving the same treatment to a capon or pullet. One of his recipes makes my mouth water: a tart with pullet. It's what we would call a pot pie—meat, herbs, and egg-thickened broth enclosed in pastry. Another recipe gives directions on preparing a pie with cockscombs, livers, and testicles. That one exemplifies the waste-not, want-not gusto of the day, but it makes me shudder. Here's another in the use-everything category for capon skin lasagna, which is not quite so shuddery. It contains no pasta; perhaps, the skin serves that purpose.

Take boiled capon skin, cut into pieces and put in fatty capon broth for a half hour with a bit of saffron. Then serve in bowls topped with a bit of cheese and spices.

Maestro Martino has much to say about eggs. One of his recipes will be found in the egg section of this chapter.

Giambattista della Porta offers several recipes, including suggestions for making sure that chickens are tender. His advice to frighten hens by hurling them down from a height and to bind a cock to a fig tree has already been mentioned, as have his instructions for cooking a chicken without a fire. He also says that hunting the birds with a falcon is sure to "make them so tender that it is wonderful." Herewith, his recipe for a capon that is simultaneously boiled and roasted:

Put a Capon well pulled, and his Guts taken out, into a Silver dish, and fill the one half of him with Broth, and put him into an oven. For the upper part will be roasted by the heat of the oven, and the under part will be boiled. Nor will it be less pleasant to behold.

It's Ulisse Aldrovandi who gives us instructions for chicken cookery that we can follow today. The man clearly savored chicken, and many pages of his chicken book are devoted to recipes—recipes for whole chickens, recipes for broth, recipes for preserving eggs, recipes for preparing eggs for the table in a dozen ways. Not a few of them come from *Apicius*. Before quoting one of these, it will be well to offer a little translation. *Liquamen* is a fish sauce (think thin Worcestershire sauce), and *defructum* is a syrup made of grape juice. Here's *Apicius* via Aldrovandi on Chicken with Milk and Pastry Sauce:

Braise the chicken in liquamen, oil, and wine, to which you add a bouquet of fresh coriander and onions. Then, when done, lift it from its stock and put into a new saucepan milk and a little salt, honey, and very little water. Set by a slow fire to warm, crumble pastry, and add gradually, stirring continually to prevent burning. Put in the chicken whole or in pieces, turn out on a serving dish, and pour over the following sauce: pepper, lovage, origan, add honey and a little defructum and cooking liquor. Mix well. Bring to the boil in a saucepan. When it boils, thicken with starch and serve.

Aldrovandi hardly depends solely on *Apicius* but comes up with recipes based on local methods. One calls for cooking chicken with grapes in a covered pot. Another is putting chicken in a pot with wine, meat broth, salt, and saffron. He gives details, as well, on how to prepare chicken parts that people in today's Western world would not consider eating: comb, wattles, and brains.

We are lucky to have an eighteenth-century English cookbook, *The Art of Cookery Made Plain and Easy*, by A Lady, as the author is identified on the title page. It was not till the 1930s that the author was discovered to be Hannah Glasse (1708–1770). The book stayed in print till about the middle of the nineteenth century. Of course, she gives chicken recipes, including this one for Chicken Hash. A sippet is a small piece of roasted or fried bread used to sop up gravy.

> To hash a Fowl. Cut it up as for eating, put it in a tossing pan, with half a pint of gravy, a tea-spoonful of lemon-pickle, a little mushroom catchup, a slice of lemon, thicken it with flour and butter; just before you dish it up, put in a spoonful of good cream; lay sippets round your dish, and serve it up.

Another eighteenth-century recipe is quoted by Dr. Kerr, who edited and greatly enlarged the Reverend Mr. Dixon's book on ornamental and domestic poultry. It comes from Chevalier Dennis de Coetlogon, a Frenchman who settled in England in 1727. Between 1741 and 1745, M. de Coetlogon published *The Universal History of the Arts and Sciences*, which presented several recipes, including this one:

> *Fowls are pickled* with Vinegar, Salt, Pepper, and Lemon-peel, and are left in their Pickle till they be wanted; when wanted, they are taken out, put to drain, and after they have been fried in Butter, they are put to stew for a few Minutes in some of the Pickle, and then carried to Table.

THE
ART
OF
COOKERY,
Made PLAIN and EASY;

Which far exceeds any THING of the Kind ever yet Published.

CONTAINING,

I. Of Roasting, Boiling, &c.

II. Of Made-Dishes.

III. Read this Chapter, and you will find how Expensive a *French* Cook's Sauce is.

IV. To make a Number of pretty little Dishes fit for a Supper, or Side-Dish, and little Corner-Dishes for a great Table; and the rest you have in the Chapter for *Lent*.

V. To dress Fish.

VI. Of Soops and Broths.

VII. Of Puddings.

VIII. Of Pies.

IX. For a Fast-Dinner, a Number of good Dishes, which you may make use for a Table at any other Time.

X. Directions for the Sick.

XI. For Captains of Ships.

XII. Of Hog's Puddings, Saufages, &c.

XIII. To Pot and Make Hams, &c.

XIV. Of Pickling.

XV. Of Making Cakes, &c.

XVI. Of Cheesecakes, Creams, Jellies, Whip Syllabubs, &c.

XVII. Of Made Wines, Brewing, *French* Bread, Muffins, &c.

XVIII. Jarring Cherries, and Preserves, &c.

XIX. To Make Anchovies, Vermicella, Ketchup, Vinegar, and to keep Artichokes, French-Beans, &c.

XX. Of Distilling.

XXI. How to Market, and the Seasons of the Year for Butcher's Meat, Poultry, Fish, Herbs, Roots, &c. and Fruit.

XXII. A certain Cure for the Bite of a Mad Dog. By Dr. *Mead*.

BY A LADY.

LONDON:

Printed for the AUTHOR; and sold at Mrs. *Ashburn's*, a China-Shop, the Corner of *Fleet-Ditch*. MDCCXLVII.

[*Price* 3 s. *stitch'd, and* 5 s. *bound.*]

at Mrs Wharton's, at the Blue-coat Boy, near the Royal Exchange

⌃ The Art of Cookery Made Plain and Easy

In the United States, it was only in the late 1800s that attention came to focus on fattening chickens. Of course, the birds had run around in villages and barnyards since people had first noticed and domesticated them. Nor were they ignored by cooks. Amelia Simmons (dates unknown), who wrote the first American cookbook, *American Cookery*, published in 1796, gives this country's first instructions on preparing chickens for the table. She tells cooks that both roosters and hens are good, "and the yellow leg'd the best, and their taste the sweetest." She advises using eyes and nose to tell if a bird is old or young, fresh-killed or "stale." "Their smell denotes their goodness; speckled rough legs denote age, while smooth legs and combs prove them young."

Martha Dandridge Custis Washington (1731–1802), the wife of George Washington,

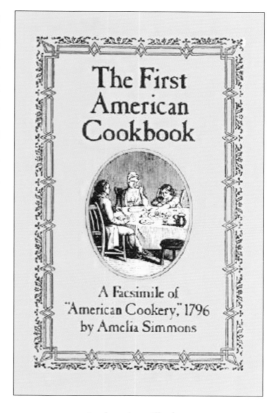

≈ *American Cookery*

may well have been familiar with Amelia Simmons's book. She herself, like many women of her time, compiled a handwritten collection of favorite recipes. Thanks to food historian Nancy Carter Crump, here is Martha Washington's recipe for a "Frykecy"—Fricassee, that is. "Flaw" means "skin"; "walme" means "bubble" or "boil."

Take a Chicken, or a hare, kill & flaw them hot, take out they' intrills & wipe them within, cut them in pieces & break theyr bones with A pestle. Y^n put halfe a pound of butter into ye frying pan, and fry it till it be browne. y^n put in ye Chiken[sic] & give it a walme or two. Y^n put in halfe a pinte of faire water well seasoned with pepper; & salt, & a little after put in a handful of parsley, & time, & an onion shread all small. fry all these together till they be enough, & when it is ready to be dished

Martha Dandridge Custis Washington »

up, put into yᵉ pan yᵉ youlks of 5 or 6 eggs, well beaten & mixed wᵗʰ A little wine vinegar or juice of leamons. Stir thes well together least it Curdle, yⁿ dish it up without any more frying.

Amelia Simmons and probably Martha Washington were dealing with everyday barnyard birds, not birds that had been raised specifically to provide good eating. The problem of how best to prepare scrawny or tough old chickens persisted for most of the next hundred years. In her cookbook *The Virginia Housewife*, which appeared in 1824, Mary Randolph (1762–1828) gives this recipe for "Soup of Any Kind of Old Fowl" and describes it "as the only way they are eatable." As you can see from her recipe, given below, she did have a notion that it might be well to do a little fattening before consigning them to the soup pot:

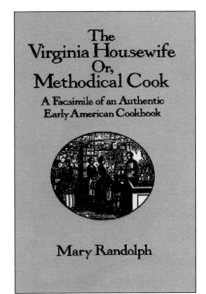

Put the fowls in a coop and feed them moderately for a fortnight; kill one and cleanse it, cut off the legs and wings, and separate the breast from the ribs, which, together with the whole back, must be thrown away, being too gross and strong for use. Take the skin and fat from the parts cut off which are also gross. Wash the pieces nicely, and put them on the fire with about a pound of bacon, a large onion chopped small,

« The Virginia Housewife

some pepper and salt, a few blades of mace, a handful of parsley cut up very fine, and two quarts of water, if it be a common fowl or duck—a turkey will require more water. Boil it gently for three hours, tie up a small bunch of thyme, and let it boil in it half an hour, then take it out. Thicken your soup with a large spoonful of butter rubbed into two of flour, the yelk [sic] of two eggs and half a pint of milk. Be careful not to let it curdle in the soup.

Dr. J. J. Kerr chimes in with his own take on tenderizing the oldest and toughest fowl:

When the Fowl is plucked and drawn, joint it as for a pie. Do not skin it. Stew it five hours in a close sauce-pan, with salt, mace, onions, or any other flavouring ingredients that may be approved: a clove of garlic may be added where it is not utterly disliked. When tender, turn it out into a deep dish so that the meat may be entirely covered with the liquor. Let it stand thus in its own jelly for a day or two (this is the grand secret); it may then be served in the shape of a curry, a hash, or a pie, and will be found little inferior to pheasant.

No danger that anyone would get *E. coli* from old fowls cooked to a Randolph or Kerr fare-thee-well.

The year 1878 produced a prodigious compendium of recipes in the form of *Housekeeping in Old Virginia*, edited by Marion Cabell Tyree, the last surviving grand-daughter of Patrick Henry. Mrs. Tyree gathered housekeeping tips and cooking instructions from 250 Virginia ladies, including Mrs. Robert E. Lee. Here's a recipe for chicken soup contributed by Mrs. P. W.:

Roasted chicken »

Put on the fire a pot with two gallons water and a ham bone, if you have it; if not, some slices of good bacon. Boil this two hours then put in the chicken and boil until done: add one-half pint milk and a little thickening, pepper and salt to the taste. After taking off the soup, put in a piece of butter size of an egg. Squirrel soup is good made the same way, but takes much longer for a squirrel to boil done.

Two more recipes gathered by Mrs. Tyree will be found in the section devoted to egg cookery.

⌃ Toulouse-Lautrec

We find a novel means of tenderizing chicken in *The Art of Cuisine* by the Post-Impressionist painter and poster-maker Henri de Toulouse-Lautrec (1864–1901), whose recipes were edited after his death by his lifelong friend Maurice Joyant. The cookbook, now a collector's item, is gloriously illustrated throughout with Toulouse-Lautrec's paintings and drawings. To make chicken succulent à la Toulouse-Lautrec, here are his instructions, which are reminiscent of tenderizing methods suggested by della Porta:

In order to make chickens immediately edible, take them out of the hen-run, pursue them into the open country, and when you have made them run, kill them with a gun loaded with very small shot.

The meat of the chicken, gripped with fright, will become tender. This method used in the country of the Fangs (Gabon) seems infallible even for the oldest and toughest hens.

It probably need not be said that, far from tenderizing an animal, fright toughens it. The reason is that the body of a scared animal or bird releases adrenaline, a hormone that leads to unchewability.

Toulouse-Lautrec also offers several mouthwatering recipes, including this one for *Fricassee de Poulet*:

Have a nice chicken and joint it.

In a cast-iron cooking pot, before a good wood fire, cook lightly some pieces of bacon. When they are nicely golden, take them out and lightly cook the pieces of chicken. Mix them all up and sprinkle with flour and generously with shallots finely chopped with parsley. Salt, pepper, and cover. Let it simmer slowly and for a long time on a gentle heat. Add a good glass of cognac half an hour before serving.

A modern cookbook writer, Barbara Kafka, has appended notes on the ingredients: "¼ lb. bacon, 1 Tbs. flour, 5 shallots. Cook 2 hrs. Add 1/3 cup cognac." But, from my point of view, the more cognac, the better. I'd put in at least ½ cup for that nice chicken.

And here's Toulouse-Lautrec on Chicken Marengo, a savory dish first prepared of locally available ingredients for Napoleon Bonaparte in 1800 after he defeated the Austrians in the Battle of Marengo. (It's sometimes called a savory Italian dish, but Napoleon's chef was a Frenchman named Dunand, who also added prawns to the dish he concocted for that momentous occasion. Only by happenstance was it made in Italy.)

Put in a saucepan some olive oil, a crushed clove of garlic; heat and brown pieces of chicken.

When these pieces are a good golden color, take them out and make a *roux* with a spoon of flour.

When the *roux* is well browned, moisten with good bouillon, put back the pieces of chicken, salt, and pepper, and let simmer on a low flame.

Half an hour before serving, add some sautéed mushrooms, a few spoons of tomato

Chicken Marengo »

purée, and pitted olives. Just as you serve, sprinkle with croutons of bread fried in butter.

Ms. Kafka's notes read as follows: "2 Tbs. oil, 1 Tbs. flour, 1 cup chicken stock, ½ lb. mushrooms sliced and heated in 2 Tbs. butter, 2 Tbs. tomato purée and ¼ cup black olives."

I think that a few more notes are needed to make the recipe not just workable but deliciously so: 3 boneless, skinless chicken breasts cut into 1-inch pieces; 1 medium onion, diced and sautéed with the mushrooms; and 1 14.5-ounce can diced tomatoes with the juice. Instead of 1 cup chicken stock, use ½ cup chicken stock and ½ cup dry white wine.

Herewith, chicken recipes that are popular at my favorite restaurant—my kitchen, that is.

Chicken Tarragon

Infinite are the ways in which to cook chicken, and here's one of my family's all-time favorites, a dish that my grown children request when they come to visit. The tarragon, a perennial herb, grows in my garden. It can be used fresh or dried. I dry a lot because it is exceedingly pricey at the grocery store. The best accompaniment to this dish is rice with tarragon-and-onion-flavored chicken broth poured over it.

Ingredients

3 tablespoons butter

1 chicken cut into pieces (breasts, drumsticks, thighs, wings, backbone, with skins and bones intact)

1 large onion, diced (more if you wish)

1 heaping teaspoon dried tarragon (or 1 tablespoon fresh leaves)

2 chicken bouillon cubes

Water

Melt the butter over medium heat in a large skillet. Place the chicken pieces in the skillet, skin-side down, and brown them briefly. Turn them over. Add the onions, the tarragon, and the bouillon cubes. Add enough water to come halfway up the sides of the chicken pieces. Bring to a boil.

Cover the skillet and turn the heat to low. Simmer for 45 minutes, checking after 25 minutes to see if a little more water is needed.

Serves 6. Freezes well.

COQ AU VIN

Many recipes for coq au vin exist, some more elaborate than others. Here's one of the fancier sort. Its deliciousness is guaranteed.

Ingredients

4 bacon slices, chopped

4 skinless, boneless chicken breasts

Salt and pepper to taste

8 ounces baby bella mushrooms, halved

1 large onion, chopped fine

2 cloves garlic, minced

1½ cups dry red wine

1½ cups chicken broth, divided

4 teaspoons all-purpose flour

Preheat oven to 300°F.

Sauté bacon in a non-stick skillet over medium high heat until crisp. Using a slotted spatula, transfer to a bowl.

Sprinkle chicken breasts with salt and pepper. Add them to the drippings in the skillet. Sauté until cooked through, about 6 minutes on each side. Transfer to a pie dish. Place in the oven to keep warm.

⌃ Coq au vin

Add mushrooms and chopped onion to the skillet. Sprinkle them lightly with salt and pepper. Sauté until brown, about 4 minutes. Add garlic; toss 10 seconds. Add wine and 1¼ cups broth. Bring to a boil, stirring occasionally. Boil 10 minutes.

Meanwhile, place flour in a small cup. Add ¼ cup broth, stirring until smooth.

Add flour mixture to the sauce. Cook until the sauce thickens, 3 to 4 minutes. Season with salt and pepper.

Add chicken to the sauce. Stir and serve.

Serves 4 to 6, depending on the size of the breasts.

Roasted Chicken Thighs

Chicken thighs roast deliciously. What's more, they are an economical cut. I buy packs of ten. The cost comes out to a little more than forty cents each. Serve the thighs with mashed potatoes or microwaved Japanese noodles, plus salad.

Ingredients

2 tablespoons olive oil

4 chicken thighs

1 large onion, sliced into four rounds

¾ cup water

salt, pepper, and paprika (adobo powder may be used instead of these)

Preheat oven to 425°F.

Put the olive oil in a 9x11-inch glass baking dish. Coat the thighs on both sides. Set them on a plate. Put the onion rounds into the dish, and place the chicken thighs

Roasted chicken thighs »

atop them. Pour water into the baking dish. Sprinkle the thighs with salt, pepper, and paprika or with adobo powder.

Bake for 50 minutes.

Serves 4 or 5, depending on appetites.

CHICKEN BREASTS IN HERBED TOMATO SAUCE

If ever a delicious recipe took no effort on the part of the cook to achieve perfection, this is it. Serve it over egg noodles or rice.

Ingredients

3 tablespoons butter, room temperature

1 garlic clove, minced

1 teaspoon dried marjoram

⌄ Chicken breasts in herbed tomato sauce

½ teaspoon sweet paprika

Kosher salt and freshly ground black pepper

2 boneless, skinless chicken breasts

10 ounces grape tomatoes

Blend butter, garlic, marjoram, and paprika in a small bowl. Season the marjoram butter with kosher salt and pepper to taste.

Melt 1½ tablespoons marjoram butter in a medium heavy skillet over medium heat. Season chicken with salt and pepper. Add chicken to the skillet, cover, and cook until no longer pink in the center, about 5 minutes per side. Transfer chicken to a plate.

Increase heat to high. Add tomatoes to the skillet and cook, stirring occasionally until they begin to char and burst, about 5 minutes. Add remaining marjoram butter to the skillet. Crush tomatoes slightly to release juices. Stir 1 minute. Season sauce to taste with salt and pepper. Spoon sauce over chicken. (Or add chicken to sauce in the skillet.)

Serves 2.

Chicken Noodle Soup Picante

What would life be without chicken noodle soup? Here's my favorite variation on that theme. It won't cure anything, but it does taste mighty good.

Ingredients

4 chicken thighs

10 cups water

1 large onion, diced

6 cloves garlic, finely chopped

1 1-inch piece of ginger, cut into 3 chunks

3 whole dried hot peppers (preferably cayenne) or 1 crushed

6 ounces angel hair pasta

⩒ Chicken Noodle Soup Picante

4 scallions, finely chopped

Combine the thighs, water, onion, garlic, ginger, and peppers in a large stockpot. Bring to a boil over high heat. Reduce the heat to low and simmer for 1 hour.

Remove the chicken from the pot and let it cool. Remove skin and bones. Return the cut-up meat to the pot.

Add the pasta. Cook, stirring with a fork, for 2 minutes or until the pasta is tender. Remove the ginger and peppers from the pot (unless you've used a crushed pepper). Ladle the soup into bowls and sprinkle with scallions.

Serves 6. Freezes well.

Eggs

Eggs figure in many dishes, from custards to omelets. They are unequaled as emulsifiers. And they are nutritional marvels, containing only seventy-five calories yet supplying all the amino acids and minerals that people require daily, along with a protein of the highest quality. They're also a good source of vitamins B and D.

And here's a bit of culinary biochemistry. Why do eggs congeal when they're cooked? Because eggs, and egg whites in particular, are full of protein. Proteins consist of polypeptides, which are chains of amino acids bonded together. Heat breaks the bonds, as does whisking. Best to cook eggs on medium heat, for high heat toughens them.

Eggs can be frozen, though not in the shell. Like a full bottle of milk, the contents expand when frozen, exploding the shell. To freeze eggs, crack them and empty the contents into a freezer container. Scramble them with a little salt, and seal the container. They'll last in the freezer for up to six months.

As for storing regular eggs, they'll keep well in the refrigerator for up to three weeks. To test eggs for freshness, place them in a pan of water. Eggs that sink and stay on the bottom are as fresh as can be. If the large end rises a bit, the egg is heading toward middle age. If the

⌃ Eggs and more eggs

large end stands completely upright, then the egg, filled with air at the top, is elderly and ought to be used forthwith. If it floats, toss it.

As I've mentioned, fertilized eggs are eminently edible. No embryo forms without incubation or artificial heat. Sailors on long voyages in sailing ships would, however, eat eggs containing unhatched chicks; the heat in the holds acted as an incubator. If that's all there was, that's what you ate. And in the Far East—the Philippines, Vietnam, Malaysia, China—a favorite food is *balut*, a cooked duck or chicken egg containing a well-developed embryo. The egg and its nascent duckling are hard-cooked and eaten warm.

A chapter in Maestro Martino's *The Art of Cooking* is devoted to the egg, from deep-fried eggs and eggs poached in water to eggs cooked on a spit and eggs in the shape of ravioli. Here is his recipe for eggs coddled in their shell. We'd call them soft-cooked eggs. "Place fresh eggs in cold water and boil for the time it takes you to say a Lord's Prayer, or a little longer, and remove."

Aldrovandi has a good deal to say about eggs, of course. The culinary chapter of his chicken book deals first with sucking eggs, for which he finds two methods, both dating back to classical Greek times and both used for eggs eaten at breakfast. One method consists of merely heating the egg, then mixing its still liquid contents with wine or garum, that early version of Worcestershire sauce. Aldrovandi lauds this method because it can supply food and wine at the same time. The other method amounts to poaching the egg, then dipping bread into its yolk. He cites Horace and Cicero as sources documenting the Roman custom of beginning banquets with egg appetizers. And he mentions conventional methods of cooking eggs, from frying and scrambling to soft- and hard-boiling. But with some bemusement, he reports the unconventional egg-cooking employed by Babylonian hunters, who placed a raw egg in a sling and whirled it till the egg was cooked by the motion. As for opening eggs, our Renaissance man notes several ways: Jews crack them at the small end so that if a drop of blood appears, they may refrain from eating it; Germans crack them on the side; and Italians open them on the blunt end.

Here's Aldrovandi's recipe for poached eggs, which he calls "Boiled Eggs." *Agresta* refers to sour grape juice. Porphorus is either a friend or a colleague.

> Place fresh eggs, their shells removed, into boiling water. When they have grown together or set, remove them at once. They should be soft; you can flavor them with sugar, rose water, sweet spices, *agresta*, or with the juice of an orange. Some people sprinkle them with grated cheese; neither I nor Porphorus care for this, though we often eat boiled eggs. The eggs are best and more pleasant without cheese.

His recipe for fried eggs involves mincing the yolks of hard-cooked eggs; mixing them with cheese, raisins, and herbs; pounding in egg whites; and frying the works in oil over a slow fire.

Marion Cabell Tyree's *Housekeeping in Old Virginia* features a baker's dozen of nineteenth-century American egg recipes. For reasons that I cannot guess, one, donated to Mrs. Tyree's cause by Mrs. S., is called Rumble Eggs:

Beat up three eggs with two ounces fresh butter or well washed salt butter. Add a teaspoon cream or new milk. Put all in a saucepan and stir over the fire five minutes. When it rises up, dish it immediately on toast.

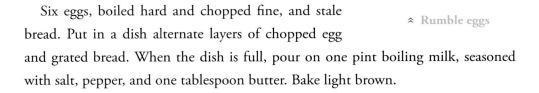

≽ Rumble eggs

And here's Miss N.'s recipe for Eggs à la Crème:

Six eggs, boiled hard and chopped fine, and stale bread. Put in a dish alternate layers of chopped egg and grated bread. When the dish is full, pour on one pint boiling milk, seasoned with salt, pepper, and one tablespoon butter. Bake light brown.

My kitchen produces some egg favorites, of course.

EGG TIMBALES

Directions for Egg Timbales, a recipe of the same vintage as Miss N.'s Eggs à la Crème, comes from my grandmother's notebook, in which she wrote her favorite recipes in an almost illegible script. In her day—she lived in or near Staunton from 1875 to 1968—Southern women cooked or supervised someone who did. My grandmother and her cohorts assumed that a list of ingredients and only minimal instructions were necessary. To suit today's cooks, I've added bracketed clarifications to her directions. If you have timbale molds, fine. If not, a muffin tin will do the job. Any leftover timbale mixture may be cooked in a small ramekin that's surrounded by hot water.

Vegetables like spinach and chopped asparagus may be added to the egg mixture. You may also use cooked fish or shrimp. Imagination can conjure many other ingredients to enhance the taste of these timbales.

It's the grated onion and cayenne that bring this dish to perfection.

Ingredients

4 eggs

½ cup milk

½ teaspoon salt

1 teaspoon grated onion

Dash cayenne pepper

Beat the eggs well. Add them to the milk. [Put milk and eggs in a saucepan and] bring them [almost] to the boiling point. Add seasonings.

Fill [greased] timbale molds [or muffin tin] 2/3 full. Put [molds or muffin tin] in a pan of [hot] water 2/3 up [to the level of the mix in the molds or muffin tin].

Bake 350°F [for 20 minutes or until a knife comes out clean].

Serves 2 to 6, depending on whether appetites are ladylike or hearty.

TOAD IN THE HOLE

The following recipe boasts many names, including gashouse eggs, one-eyed jack, bird's nest, and moon eggs. "Gashouse" sounds as if it connotes something indigestible, but the word has an honorable provenance. I've heard that it referred originally to a workman's shed on the Gowanus Canal in Brooklyn. A man could cook a simple meal in the gashouse, and what simpler than bread cradling an egg? The term was not restricted to Brooklyn: In the 1941 movie *Moon over Miami*, the father of her fiancé teaches Betty Grable how to cook gashouse eggs. The movie's title has also been used as a title for this recipe. When my children were small, the recipe was given to me as "Toad in the Hole," and so we have known it to this day. The instructions call for a single serving. Use more bread, eggs, and butter as you will.

Ingredients

1 slice bread

1 egg

3 tablespoons butter

⌃ Toad in the Hole

In a skillet, melt 1½ tablespoons butter over medium heat.

With a cookie cutter or a glass with a 2- to 3-inch diameter, cut a hole in the center of the piece of bread. Place the bread, including the hole, into the skillet. Break an egg into the hole. Fry till the bread begins to turn golden-brown and the egg to set.

Melt the remaining butter while briefly lifting the bread and egg out of the skillet.

Return the bread and egg, uncooked side down. Continue frying until the egg is cooked to your liking.

Serves 1.

POACHED EGGS

Poached eggs on toast, poached eggs on corned beef hash, poached eggs Benedict on English muffins with Canadian bacon and Hollandaise sauce—poaching brings out the egg's versatility. A conventional recipe follows, but water is far from the only medium in which the eggs may be immersed. Try chicken stock, tomato soup, wine, or stewed tomatoes. If you use any of these four, there's no need to add vinegar and salt.

Ingredients

4 eggs

Water

2 teaspoons vinegar

¼ teaspoon salt

Break the eggs into separate small bowls.

Put 2 inches of water into a large skillet, and bring it to a boil. Add the vinegar and salt. Reduce the heat to a simmer.

Add the eggs one at a time and poach for 2 to 3 minutes, until the egg whites set. Remove the eggs from the water with a slotted spoon and trim away any small shreds of egg white.

Serves 2 to 4, depending on appetites.

EGGS BENEDICT

These are the brunch eggs supreme. The recipe for Hollandaise sauce follows the recipe below. It can be made just before the eggs are poached.

Ingredients

4 eggs

2 English muffins, split

4 slices ham or Canadian bacon

Hollandaise sauce (recipe follows)

Eggs Benedict »

Poach the eggs as in the recipe above. Toast the English muffins, and heat the ham or Canadian bacon gently in a skillet. Place the meat on the muffins.

Set a poached egg atop each ham- or bacon-lined muffin. Top with Hollandaise sauce.

Serves 4.

BLENDER HOLLANDAISE SAUCE

Ingredients

1½ cups (2½ sticks) butter

2 egg yolks

2 tablespoons fresh lemon juice

Salt

Pepper, freshly ground

Fill a blender with hot water, and set it aside.

Melt the butter in a small saucepan over medium heat until foaming. Remove pan from the heat.

Drain the blender and dry it well. Put egg yolks and 2 tablespoons lemon juice in the blender. Cover and blend to combine.

With the blender running, lift the cover and slowly pour the hot butter into the blender in a thin stream of droplets. Discard the milk solids on the bottom of the pan. Blend until creamy sauce forms. Season with salt and pepper to taste.

"Smoked" Eggs

These hard-cooked eggs aren't really smoked, nor do they taste of smoke. They're dyed, and gloriously so. The shells attain a rich plum purple, while the cooked whites become an attractive pale mustard-yellow. You can use a white crayon to make designs on the shells before cooking the eggs. The result: the fanciest and tastiest possible eggs. Nor should they be reserved for a special occasion like Easter.

The tool that you'll need is a slow cooker or an oven-proof dish with a tight-fitting lid.

Ingredients

Eggs in the shell
Yellow and red onion skins, the papery part
Water
Olive oil

⩗ Smoked eggs

Place a thick layer of onion skins on the bottom of the cooker. Add the eggs. Cover with another thick layer of onion skins.

Add water to cover the eggs and skins by an inch. Drizzle oil on the water to make a thin slick. Put the lid on the cooker.

Cook at 225°F in the oven or on low in a slow cooker for at least 8 hours. Up to 16 hours is all right.

Eat the eggs as is, devil them, or use them for egg salad. Enjoy.

Gourmet Deviled Eggs

If you'd like to make the best deviled eggs ever, this recipe will give you directions. It avoids the sweet-pickle taste of most deviled eggs. Its zesty secret is the addition of an onion-rich French dressing, for which instructions are also given below. It was a favorite of my grandmother, who may have found the recipe in an issue of *Gourmet* magazine that dated back to the 1940s or before.

Ingredients

Eggs, as many as you'd like
Mayonnaise
Garlic powder
Celery salt
Dry mustard
Gourmet French Dressing (recipe follows)
Paprika
Capers (optional)

Boil the eggs until hard. Cool and shell them. Cut them lengthwise and put the yolks into a bowl.

Mash the yolks with a fork. Add a little mayonnaise at the rate of 1 tablespoon for every 6 eggs.

⌃ Gourmet deviled eggs

Add garlic powder, celery salt, and mustard. Mash in Gourmet French Dressing until the yolks are creamy and lumpless.

Fill the hollow of each egg white with the yolk mixture. Sprinkle paprika atop the yolk mixture. If you wish, top each deviled egg with a caper.

GOURMET FRENCH DRESSING

Ingredients

1 10½-ounce can tomato soup, undiluted

1 cup cider vinegar

½ cup sugar

1 cup vegetable oil

1 tablespoon salt

½ teaspoon pepper

1 tablespoon Worcestershire sauce

1 tablespoon dry mustard

Juice and pulp of 1 large onion

Mix all ingredients together in a quart jar. Refrigerate.

Bon appétit!

CHAPTER SIXTEEN

Hen Music

"Was Ubon at your house this afternoon?" Mike, my across-the-street neighbor, has succumbed to curiosity.

"She was," I say. Small, roly-poly Ubon has a loud and penetrating voice. "I bet you could hear her."

But I'm getting ahead of this part of the story, which began when the Holland lop, that bad-tempered and boring bunny, died just before Thanksgiving. The ground here is too full of baseball-sized rocks to dig a decent grave, and no deep woods are available for a sky burial. Put into a paper sack that had once contained cat kibble, she was consigned to the trash. In life, she'd been a fine producer of poop for the garden, but other than that, she'd spent her three and a half years entirely without love. At seven o'clock on the morning that the ad for the sale of the hutch appeared, the phone started ringing and continued to ring without much pause for the next three days. (Rabbits are popular!) The seven o'clock caller bought it, planning to use it as a haven for a nursing doe.

My imagination began to hear cooing, peeping, chirping, clucking. A small flock of hens—they were animals with which I could socialize, animals that would become friends, animals to love. My backyard needed them.

Before Bunny had arrived, I'd talked with Sue Randall of Elk Run Farm. She and her husband, Jim, were running their booth at our downtown farmers' market. It was early May, and I'd come to pick up Better Boy and Mortgage Lifter tomato seedlings. Sue and Jim sell not only seedlings but also veggies, goat meat, and lovely brown eggs. "What kind of chickens do you have?"

"Red sex-link," she said. "We get them when they're seventeen weeks old and ready to lay."

"And where do you get them?"

"Westdale Hatchery up in Harrisonburg."

End of conversation, for the market was crowded, as always, and Sue's booth six deep in customers. Back home, there was information to be gathered. Just where is Westdale? Does it sell chicks as well as pullets ready to lay? And what the dickens are red sex-link hens? A quick

⌃ The hutch

call solved the first two questions: only pullets, purchase in spring. As for sex-link chickens, that took a bit of investigation.

Sex-link birds are hybrids in which the cockerel chicks and the pullet chicks can be sexed by their colors at the very moment that they emerge from the egg. The physical sexing of chickens is more art than science. Roosters keep their sex organs inside their bodies. Sexing is accomplished by picking up a day-old chick, squeezing the poop out of its anal vent, and looking to see if there is—or is not—a shiny bump on the middle and front portions of the vent. A pullet is bumpless and lacks the anal shine. The bump, called a papilla, will become the mating organ when the chick matures into a rooster. This method of distinguishing male from female was first discussed in a paper published in Japan in 1933. One episode of the Discovery Channel's *Dirty Jobs* features host and dirty-jobber Mike Rowe at the Murray McMurray Hatchery where he's trying his hand at sexing chicks. Squeeze chick and hope that the yellow feces land in what looks like an otherwise empty three-pound coffee can. A skilled sexer can achieve as much as 95 percent accuracy. Nonetheless, some roosters will slip through. It was just such a slip that brought Sally, the "accidental" rooster, into Lilly's flock. A much simpler method of sexing is simply to let your day-old chicks grow up a little. It takes only six weeks or so for a rooster to start looking like a rooster with nascent spurs, incipient wattles, prominent comb, and elongating neck hackles. The unwanted roosters can be raised for meat, and hens kept for both chumminess and eggs.

And then there's sex-linking, a method that allows for no accidental roosters whatsoever. With sex-link chickens, you can tell the cockerels from the pullets by the color of the down when they hatch—females will have one coloration, males another. Hybridization confers this gift. The secret, which was likely put into play sometime in the 1990s, lies in using sex-linked color traits. In chicken genetics, mixing a barred hen with a non-barred male results in barred cockerels and plain pullets. With black sex-links, mating a Rhode Island Red or New Hampshire rooster with a Barred Rock hen equals female chicks that are black with red mixed into their neck feathers and male chicks with faded bars and light-colored neck feathers. These hybrids are also known as Black Stars or Black Rocks. With red sex-links,

⌄ Red sex-link pullet

Rhode Island Red or New Hampshire roosters plus White Rock, Silver Laced Wyandotte, or Delaware hens equal silvery males and red or buff females. Like the black sex-links, they have been given fancier names, like Red Star. Red Star x Red Star will not produce sex-link chicks. These hybrids do not breed true. To get sex-link birds, you start afresh every time.

≫ Gold Legbar rooster

Behold! Autosexing varieties of chicken have been developed. Not only do male and female chicks hatch showing distinctively different colors, but the birds also breed true. One such breed is the Legbar, developed in the 1930s from Brown Leghorn and Barred Rock with a smidge of Araucana thrown in for good measure. The Araucana genes are responsible for the color of Legbar eggs—blue and, occasionally, olive green. Another autosexing chicken is the California White, which is a cross between a White Leghorn hen and a California Gray rooster. The male chicks are pale yellow, while the females sport black speckles on their yellow down. The California Whites, which lay white eggs, are said by some to be flighty and, by others, steady.

I also immersed myself in chicken trivia. Worldwide, chickens are by far the most common bird and, at 10 billion, there are more of them than any other bird. Hens lay some 700 billion eggs every year. Sixty recognized breeds exist (but I'm sure that there are more mongrels on earth than purebreds). An average life span runs to seven or eight years, though some chickens have managed to arrive at the supreme decrepitude of fourteen years. Chickens can achieve a speed of nine miles per hour (I've seen them run as fast as cheetahs; they just don't maintain racing speed for more than a few seconds). The longest recorded chicken flight lasted thirteen seconds and covered a distance of 301.5 feet. Not one of the websites tells how high off the ground the bird flew, nor are we told if the record-setting bird was a hen or a rooster.

Finally, I made up my mind: red sex-link chickens and a coop to house four or five hens. I went to visit Sue and Jim Randall at Elk Run, their thirty-acre farm. Sue's red sex-link hens, which come in colors from Rhode Island Red to wheaten, are enclosed in a very large moveable pen with wire fencing. Kale grew there last year, and the now-dry stalks poke up from the otherwise bare soil. Despite their pen, the hens are deemed free-range because they have ready access from their coop to the outside world. The birds look uncountable, but Sue said that she has one hundred and fifty, maybe one hundred and sixty. In warmer weather, they produce a hundred and forty eggs a day. She takes the eggs to several farmers' markets in cartons with labels announcing that the eggs were laid by "Happy Hens." She raises the birds in two-year cycles, with one batch wearing a leg band, the other unbanded. This way she knows which hens are getting too old to lay; they'll be sold as meat birds. Last year, Jim built her a feather-plucking machine.

As we approached, they skittered away from us and headed toward the large coop that contains their nest boxes. But when Sue gave a loud yodel, they skittered, cooing and clucking, back in our direction.

≽ Sue's red sex-link hens

"What happened?" I asked.

"Oh, that's the call I give when I'm bringing weeds. They love chickweed."

I loved hearing the cooing and clucking. Hen music was on its way to my yard. I emailed AnneMarie Cumiskey, the coop lady. She called me almost immediately. I ordered one of her sturdy Douglas fir coops and made arrangements to fetch it on the day that she brought her young'uns to Charlottesville for a Thursday class. My next-door neighbor, owner of a truck, was willing to take me the thirty-five miles over the mountain to fetch it.

Fate has a way of rearranging the best-laid plans. On a January Wednesday, the day before I was to fetch the coop and take it home, I had lunch with friends at Ubon's Thai Victorian Restaurant, so named because it's located in a huge Victorian house built in the 1880s. It's located only three houses down, a right-hand turn, and three houses up from my place—easy to get to, aside from which Ubon's cooking, her fresh spring rolls, pork dumplings, pad Thai,

drunken noodles, and all the rest, is superlatively tasty. With chickens in the forefront of my mind, I told my companions about the coop-to-be. Ubon heard the word "chickens." Wearing a floor-length pink outfit resplendent with gold threads, she descended upon us.

"Chicken, chicken," she exclaimed in her resonant rasp. "Honey, I got chicken." But she was afraid that she would have to sell her chickens, for they were at her other restaurant forty miles away. Her son, Tom, was taking care of them, but he was soon to move to Staunton. Ubon feared

Ubon »

that she would lose them, all fourteen of them. She could not move them to the Thai Victorian's spacious grounds because it was zoned commercial.

Lightning struck. "Ubon, come see my yard. I'd love to have chickens there."

"Fourteen chicken," she said. "Brown egg but one lay white."

"Sounds good to me."

As soon as I got home, I called AnneMarie to tell her that strange things had happened and I wouldn't be picking up the coop. She was calm and congratulatory when she heard that I'd be acquiring a flock much too big for it.

When Ubon saw the yard, she exclaimed in her gravelly voice, "I talk to Buddha, I talk to Jesus, my prayer answered! I can see from restaurant parking lot." And she gave me a big Ubon hug.

Her son Tom came the next day to see the yard and decide on a location for the coop. Late that afternoon, he came bearing a plastic box that contained so much Thai dinner that it lasted me for two nights. Then nothing happened, and still nothing happened. I stopped by Ubon's place for takeout several times, and she assured me that chickens would be on their way to Staunton. "When I get chicken," she said, "you come." But she didn't say just when that would be.

In mid-March, though, she handed me a dozen brown eggs and a harbinger of chickens-to-be—dozens on dozens of egg cartons, some with room for twelve eggs and others with room for eighteen or twenty-four. Springtime and fourteen chickens make for a superabundance of eggs.

On the first full day of spring, I heard a hard knock-knock-knock on my kitchen door. Ubon. With her was a young man, who was up on my terrace, the place that Ubon's son Tom had chosen as an ideal place for a coop. But it's not ideal, not in the least. I am too creaky to manage the terrace's steep incline. I want the chickens where I can tend to them, feed them, collect eggs, and scoop out the poop. I want them where I can enjoy them. The young man scrambled down the steps when I came into the yard.

"This Matt. He good boy," Ubon said. "He build coop."

I pointed to the backyard spot that had held tomatoes last year and butternut squash the year before. "I can't get up there. Please build it here."

With that, Matt measured a 12 x 12-foot space and set to work with a posthole digger. He looked at me with a woebegone expression. "I'm a painter," he said. "This coop thing—I don't know what I'm doing."

Tom showed up shortly thereafter. A three-way discussion ensued. Tom insisted on a terrace location because the concrete wall up there could serve as one side of the coop. I insisted that it be built where I could get at it and where it had the proper south-facing classical orientation recommended by Pliny and his cohorts. Matt, bless him, took my side. Shortly thereafter, Tom disappeared. Later Dan Herlong, Ubon's American husband, arrived on the scene. He said that Tom had been exceedingly displeased to have his plans for locating the coop overturned. Tom was therefore sulking. I learned, too, that Tom was not the person who cared for the hens, who fed them, made sure that they had water, and cleaned their

coop. Dan himself was the hen-keeper. With Dan in charge—and eager to hand his duties over to me—construction began all over again, this time on a much more modest scale. By the end of the day, the coop had not only been framed but was capped by a corrugated roof of dull red Fiberglas. Dan was back at work the next morning and the day after. He fashioned a door big enough so that a person could enter without stooping; he made a raised floor and enclosed it so that the hens would have a safe and cozy place for roosting. To this he added three nest boxes. All that was left to be done was digging a trench for burying chicken wire around the

Dan roofing the coop »

The coop, facing south »

base of the coop and then stapling the wire to the studs. It would need a coat of wood-preserving stain, but there was no hurry for that. Then thunderstorms struck. Dan kept working until he was sopped, but the downpour and darkness put an end to his endeavors. The next day, he had to be on duty at Ubon's Thai restaurant in Nellysford, the home of the chickens.

The next day dawned dry and sunny, except for the mud underfoot. Another one of Ubon's good boys—this one twenty-seven years old—swaddled the coop in chicken wire. My son arrived and installed two roosting perches. I purchased a wood-preserving stain, which the good boy applied. We rigged a chicken ladder out of scrap wood so that the hens could easily walk up into the enclosure that held their nest boxes and would also give them shelter from wet or snowy weather. I lined the three nest boxes with straw. And none too soon.

On March 29, two days later, I returned from errands to find Dan's pickup truck parked in my driveway. The hens had been packed into a large metal cage. Dan put it on my wheelbarrow—my red wheelbarrow with a yellow tire—and pushed it over to the coop. It took the birds, which were a bit rocky from their forty-mile road trip, several minutes to realize that they were free to leave the cage. Dan installed their feeder and waterer. I took photos. The birds were skittish in their strange new home with strange new me much too close to them. I knew that they numbered fourteen and that they came in three breeds—Barred Rock,

Rhode Island Red, and Leghorn. Because they hid under the floor of the nest box chamber, I could count only the last—the lone white bird with a large and floppy Leghorn comb. But hen music began right away. Because of early spring chills, my kitchen windows were closed, but the soft, conversational music came right through the window glass.

Several sessions of loud clucking—*puck-puck-puck- p'cawk*—also made their way inside. Within three hours of arriving in their new home, the birds had laid eight eggs. Two fifty-pound bags of laying pellets were delivered that afternoon by my brother, who runs a feed mill. The next morning, a five-gallon bucket of kitchen scraps—lettuce, carrot peelings, broccoli stems, and more—sat beside the door of the coop: leavings from Ubon's restaurant. When the contents were emptied onto the coop floor, the hens dived right in, cooing and clucking softly as they ate. Omnivores, they have cunning beaks, equipped to shred just about anything. They no longer skittered when I hove into sight. Rather, they clustered at the gate, waiting to see what laying pellets or kitchen leavings might come their way. Like the

≽ Barred Rocks, Rhode Island Reds, and Gertrude

hen that knocked on E. V.'s door, they're not dumb clucks. They allowed themselves to be counted: six Barred Rocks, seven Rhode Island Reds, one Leghorn. The *American Standard of Perfection* would find all of them to be mongrels. Some of the Rhode Island Reds, though they sport rusty-red feathers, are probably hybrids, for several have white in their neck feathers and creamy or white rump feathers. Two of them are a dark, almost iridescent red. It's they, I think, who are responsible for the milk-chocolate colored eggs. A hen of a somewhat rustier red wears black spangles like sequins on her neck feathers. And then there's Gertrude, the Leghorn with a large, floppy red single comb, who stands out amid the crowd. She was the first of the hens to be named because she is so distinctively herself. She, too, would be disqualified by the APA because her legs are cream-colored rather than the requisite yellow.

What would Varro, Columella, and Aldrovandi think of these ladies? I'm sure that they'd approve of thirteen of these sturdy, big-breasted hens. Only Gertrude would meet with disapprobation because she is white and, in their experience, white hens are not only poor layers but also short-lived. They had never met the prolific and hardy Leghorn, though that's an Italian breed. They might wonder, too, why the flock has no rooster, an absence that I'm not sure I could explain to them. The coop, facing south as Columella stipulated, would receive high marks. Their shades, at home now in the Elysian Fields, are pleased.

Dan told me that it would be best not to let the chickens out into the yard until they knew for certain where home and safety lay. Much of the backyard consists of raised beds and in-earth planting plots. Chickens and gardens do not agree. I'd thought that netting tents and impromptu chicken-wire fences would suffice to protect the carrots, spinach, beans, cukes, and garlic. The tomato cages could be shielded in the wire left over from building the coop. But a book on chicken-keeping gave me warning:

> Chickens are beautiful. Chickens are fun. Chickens are also a bit rambunctious. Left to run unattended in your garden, gentle hens take on the demeanor of roadhouse thugs. They break blossoms. They crush tender shoots. They pull up baby lettuce and lay siege to unsuspecting squash seedlings. They don't mean to; they're just a band of happy, clumsy hens.

⌃ The gated and fenced hen yard

Rethinking was in order. It dictated building a new gated chicken run, surrounded by a chain-link fence, that would allow the ladies room to roam outside in search of greens, insects, and worms. Among much else, that part of the yard offers them redroot pigweed, dandelions, grass, and an unlimited number of harlequin bugs.

Within days, the chain-link run was put in place. The ladies came out to dust-bathe, peck, and converse. I learn that I can call them "free-range" birds because they are not confined to their coop. They begin to find names, though the pessimistic guys who wrote *The Chicken Book* have instructed me not to become so intimate with them:

Chickens are not pets; they are chickens; they are producers; they exist to lay eggs and to be eaten. Never name a chicken. To do this is merely cute—and silly—and an abuse of names. This does not mean that you cannot enjoy, admire, and love chickens

⌃ The ladies

individually and collectively; it just means that you cannot sentimentalize and falsify your relationship to chickens.

In their view, the relationship is supposed to end with the death of the bird. We grieve, sometimes inordinately, over the deaths of our pets. Therefore, do not turn hens and roosters into pets, Q.E.D. But most of my chicken people have named their birds—Oberon, Izzy, Christmas. I've also come across some wonderfully inventive names—Charlotte Brooooooonte, Jane AustHen, and Emily Chickenson. For my egg-producing ladies, names well up simply, unstoppably from a deep reservoir in my imagination. It's easier to name the hens than to speak of the white one or the Barred Rock with only one eye or the red hen with black spangles on her neck feathers. They have become, respectively, Gertrude, Una, and Hortense. Of the two dark red hens, she of the bright red comb is Helenka; her near look-alike, the bird with the dull red comb, is Hetty. Red Veronica's neck-feather spangles are white, while Permelia's cape is flecked throughout with tiny white feathers. The Barred Rock with a large, pert Dominique tail, is, of course, Dominique. The Barred Rock at the bottom of the pecking order—her breast is a bit bare—is Cinderella, victim of her mean stepsisters. Nine names down, five to go. The birds are entertaining—Gertrude picks up a worm and all the others chase her helter-skelter around and around the yard. They are not, however, pets. I do not pick them up and cuddle them. I see to it that water and food are available, that the coop is clean, its straw replaced weekly. I shove them off the nests to collect their lovely eggs.

Chickens have an undeniable social awareness. Not only do they talk and play, but a flock establishes a hierarchy, in which one bird, occupying the summit of social influence, bosses all the others, while another bird finds itself at the nadir, with the rest of the flock in the peck-and-be-pecked middle. A pecking order comes into being when the members of a community are kept in close proximity. It's a form of government, like office politics, that is designed to let individuals know exactly where they stand in the scheme of things. The ways in which chickens decide who is top bird, bottom bird, and birds between is a mystery but may well be analogous to the ways in which people establish rankings. Getting to the top requires strength, speed, agility, and wit. No hierarchies, however, are immutable. Falling from the pinnacle may involve aging, injury, death, or the entry into the group of newcomers.

When the hens arrived, the bottom of the pecking order in my small flock was Cinderella, a Barred Rock with a large bare spot in the middle of her plump breast. Other hens had pecked her featherless. The top bird was white Gertrude, the eldest and the quickest to grab a bug or a piece of bread. In only a matter of months, the order changed dramatically. Both age and injury seem to have dictated Gertrude's downfall. She lingers alone in the coop when I release the ladies each morning, and she limps.

Now that it's spring, now that the peas, carrots, and spinach are up, now that the pear trees are in bloom, the hens give me not just hen music but also ten, eleven, twelve eggs every day. The eggs are beautiful, with colors ranging from Gertrude's white through light buff, beige, rich brown, milk chocolate, and might-as-well-be purple. My phone rings, people knock on the door, the eggs sell like the proverbial hotcakes. One of the door-knockers is a slim, scruffily bearded young man, whose girlfriend works in Ubon's restaurant. He's not after eggs, though. He speaks good English with a strong, unidentifiable accent. I ask him where he's from. Saudi Arabia, and he tells me that he's grievously disappointed that places like Wal-Mart and Lowe's don't carry composted chicken litter. May he please have some poop? Given the go-ahead, he climbs right into the coop with a spoon and a plastic bag. He's invited to come back whenever he's in need.

My life has changed. It's not just that hen music makes daylight sweeter, that the ladies need a never-ending supply of food and water, that the coop and yard need clean straw. I'm cooking for

⌃ Eggs from my hens

chickens. Kitchen scraps like watermelon rinds and large carrot tops are cut into beak-sized pieces because big chunks tend to linger. I nuke the sprouting potatoes that Ubon gives us to do away with potato toxins. A friend who's heard about the hens writes to tell me that chicken-tending is *work*. Yes, but work that has a purpose; work that brings eggs and soft hen-talk doesn't really qualify as work.

Then, in May, we hold a blessing—Ubon calls it a "bressing"—of the chickens.

YOU ARE INVITED TO

THE BLESSING OF THE HENS

4 PM, WEDNESDAY, MAY 11

210 NORTH MADISON STREET

THE REV. GLENN BARRETT, OFFICIANT

REFRESHMENTS

RAIN DATE: 4 PM, MAY 11

RSVP: 886-4180

As a longtime member of my exercise class at the YMCA, Glenn, a minister now semi-retired, seemed the right person to conduct the blessing.

⤫ The blessing of the hens

St. Francis and the wolf of Gubbio »

On May 11, clouds came and went. At 4:00 PM the sun shone, and the temperature reached a balmy seventy-five degrees. The backyard filled with close to forty celebrants, including Ubon. Though I'd thought that the crowd would frighten them, the hens didn't hide. Making gentle conversation, they pecked and scratched through their yard.

Glenn began with a gentle sermon about St. Francis, including a favorite

The blessed hens

story that has nothing to do with hens but illustrates saintly empathy with the world's creatures. A wolf was not only terrorizing the people of Gubbio but also killing and eating them. Francis told them that the wolf was plainly hungry and that the animal should be fed. End of killings. Glenn read from Genesis about the creation of fish and birds. All of us participated in responsive readings and concluded with a prayer. (For anyone who'd like to conduct a similar blessing, the short, sweet service is given in the appendix.)

The human contingent then trooped inside for iced tea, wine, cookies, and other refreshments, including a big pan of tiny transparent noodles and vegetables in a special Thai sauce brought by Ubon. Someone murmured the irresistible pun: "Really a fowl blessing." Glenn received an honorarium of a dozen beige, brown, and lavender eggs. The last celebrants left at 5:30. The smiles hadn't stopped.

We are happy. So are the hens.

The Blessing of the Hens

Portions in boldface are those spoken in unison by the assembled company.

Welcome remarks and short, sweet sermon on St. Francis.

Reading by the officiant from Scripture: the creation of fish and birds, Genesis 1:20–23:

Then God said, "Let the waters abound with an abundance of living creatures, and let birds fly above the earth across the firmament of the heavens." So God created great sea creatures and every living thing that moves, with which the waters abounded, according to their kind, and every winged bird according to its kind. And God saw that it was good. And God blessed them, saying, "Be fruitful and multiply, and fill the waters in the seas, and let birds multiply on the earth."

Responsive reading of Psalm 148:3–11:

Praise God, sun and moon,
Praise God, all you shining stars.
   Praise God, you highest heavens,

and you waters above the heavens.
Praise the Lord from the earth,
you sea monsters and all deeps,
    fire and hail, snow and frost,
    stormy wind fulfilling God's command.
Mountains and all hills,
fruit trees and all cedars,
    Beasts of the earth and all cattle,
    creeping things and flying birds!
Kings of the earth and all peoples,
princes and all rulers of the earth!
    Young men and maidens together,
    old men and children!
Let them praise the name of the Lord,
    for God's name alone is exalted,
    God's glory is above earth and heaven.

Spoken hymn: "All Things Bright and Beautiful," Cecil Francis Alexander:

All things bright and beautiful,
All creatures great and small,
All things wise and wonderful:
The Lord God made them all.

Each little flower that opens,
Each little bird that sings—
God made their glowing colors,
God made their tiny wings.

Closing prayer:

Blessed are you, Lord God, maker of all living creatures. You called forth fish in the sea, birds in the air, and animals on the land. You inspired St. Francis to call all of them his brothers and sisters. We ask you to bless these hens. By the power of your love enable them to live according to your plan. May we always praise you for your beauty in creation.

Blessed are you, Lord God, in all your creatures. AMEN.

Notes

Chicken Dreams

2 "The chicken coop is the new": *The New Yorker,* February 2009.

6 "It shall be unlawful": Staunton City Code.

The Ur-Chicken

10 "Although these scaly": Zimmer, 38–39.

13 "Prior to that": How Stuff Works.

The Classical Chicken

18 "Eggs are hatched": Aristotle, 1.vi.

19 "Here hatch"; "and upon them egges"; three inches thick: St. John, 327–328.

20 A good woman: Aesop.

20 such eggs: Aristotle, VI.2.

20 "Of my own accord": Theognis, 863-864.

21 Son of Philanor: Pindar, O.XII, 19-23.

21 "Pigs wash": Heraclitus, 28.

22 ""the young bloods"; Lazenby.

22-23 "Crito, I owe a cock": Plato, Phaedo, 118a.

23 Some birds couple: Aristotle, Book VI.1.

24   About the twentieth: Aristotle, Book VI.2.

24   if you open the egg: Aristotle, Book VI, 3.

26   One could see: Smith and Daniel, 73–74.

27   "The rooster seems": Lind, 199.

27   Lo, the raving lions: Lucretius, *De rerum natura,* Book IV, ll. 710 ff.

28-29   He advises: Varro, III.ix, 3-4.

30   "After laying": Pliny, Book X.LVII, page 367.

30-31   "night-watchman"; These birds control: Pliny, Book X.LVII, page 323.

31-32   Whatever both nature: Lind, 212.

33   Let your brood hens: Columella, VIII.ii.5, 327.

33   These male birds: Columella, VIII.ii.11, 329.

34   Put grass: Columella, VIII.V.12.

34   coop design" Columella, VIII.III.1-7.

The Medieval Chicken

35   The cock: Trevisa quoted by the Oxford English Dictionary under the definition for "roost."

37-38   "The vanishing": Slavin, 54.

The Renaissance Chicken

40   There are two kinds: della Porta, 33.

40   "Mice"; "Red Toads": della Porta, 34.

40   "if the Egg": della Porta, 54.

40   "a hollow vessel"; "I tried this": della Porta, 143.

41   "drowned him alive"; Wild cocks bound: della Porta, 288.

41   "cramming": della Porta, 290.

41   But a Cock: della Porta, 144.

42   Put a piece of Steel: della Porta, 295.

42   "It is clear to all": Lind, 4.

44   They follow their chicks: Lind, 142–143.

45   I pass over now: Lind, 98.

46   "lay the best": Lind, 399.

46-47   Among the tame: Lind, 37.

The Medicinal Chicken

49   "The genus of chicken": Lind, 259.

49   "tempers the harmful": Lind, 261.

50   Throw away the entrails: Lind, 263–64.

50   Grind up eggshells: Lind, 269.

51   "Burn up": Galen quoted by Aldrovandi, Lind, 268.

51   "strength is so great": Lind, 341.

51   "Those who decorate"; "Once egg white": Lind, 343.

51   "promises that this mixture": Lind, 259.

52   Take a raw egg: Empiricus quoted by Aldrovandi, Lind, 274.

53   Three Spoonfuls: Empiricus quoted by Aldrovandi, Lind, 272.

54   has dried: Lind, 270.

The Transitional Chicken

56   "chicken breeding represents": International Chicken Genome Sequencing Consortium, 712.

59   The history: Dixon, Ornamental and Domestic Poultry, 12.

59   "deservedly the acknowledged pattern": Dixon, A Treatise, 49

59   The courage of the Cock: Dixon, A Treatise, 47.

60   Cock-broth; *The Dung;* the Weasand": Willughby quoted by Dixon, A Treatise, 45.

60-61   "the most treasured pets"; We have advanced: Dixon, A Treatise, 317.

61   THE SHAKEBAG FOWL: Dixon, A Treatise, 272–273.

62   "Meanwhile, we": Dixon, Ornamental and Domestic Poultry, xv.

62   "the several kinds of Shanghaes": Dixon, A Treatise, 10.

62   "the active enterprises": Dixon, A Treatise, 27.

62   Those authors who: Dixon, A Treatise, 27–28

63   My own belief: Dixon, A Treatise, 3.

63   "on islands"; Dixon, A Treatise, 42.

64   "Never in the history": Burnham, 9.

64   "The Hen Trade": Burnham, 325.

The Modern Chicken

66   "The "Rump-less fowls": Darwin, 230.

66   We see that: Darwin, 230.

67   "the black stately Spanish": Darwin, 231.

67   The title: Tegetmeier, 235.

68   Bantams are great: Hobson and Lewis, 40.

68   "BANTAM BREED": Darwin, 230.

69   "dwarf fowl": Darwin, 247.

69   "little whipper-snapper": Dixon, A Treatise, 321.

69   "The male and female": Darwin, 252.

70   "A much smaller race": Dixon, A Treatise, 465.

72   "Which just goes to show": MacDonald, 287.

72   "The time has passed": Smith and Daniel, 242.

76   "water, soil, insects": Bad Bug Book, entry under *Salmonella* spp.

84   "If you tell"; "A whole chicken": Salatin, 373.

84   Unrestrained technology: Smith and Daniel, 300.

The Conquering Chicken

94   "Next to the Dog," "But the most mysterious": Dixon, A Treatise, 40–41.

96   "People mostly rear them": Tsudzuki.

Eggs

106-107   began upon the following occasion: Swift, 62-63

107-108   Eggs are popularly: Dixon, A Treatise, 100.

108   "Eggs are the superfluity": Dixon, A Treatise, 109.

111   The hen announces: Lind, 99.

The Scientific Chicken

115   "The domesticated chicken": Trans-NIH Gallus Initiative.

116   "These results provide": Kannampilly.

121   "a pharmaceutical bioreactor": Alper, 729.

122–123   "Although there are": Lillico, 192.

128   "The chicken has been used": Burt, 1465.

129–130   "The chicken has been used": Burt, 1465.

129   Now that the human genome sequence: Schmutz and Grimwood, 680.

130   "For nearly every aspect": International Chicken Genome Sequencing Consortium, 712.

The Storied Chicken

135-137   I had a little hen; Chook, chook; Hickety, pickety: Opie, 201–202.

137   Cock crows, Opie, 126.

138-139   And in the yard: Chaucer, 265–266.

140   And so befell: Chaucer, 276–277.

140   This Chauntecleer: Chaucer, 278.

141   "In spite of you": Chaucer, 280.

141   But, truth: Chaucer, 281.

141-142   HORATIO: And then it started: Hamlet, Act I, Scene I.

Chicken People

148   "and exchanged them": Moses, 63.

Chicken Cuisine

166   Capon with Herbs; Chicken with Cumin: Taillevent

168   Take boiled capon skin: Maestro Martino, 71.

168   "make them so tender": della Porta, 287.

168   Put a Capon: della Porta, 293.

168–169   Braise the chicken: Lind, 318.

169   To hash a Fowl: need credit

169   Fowls are pickled: Dixon, A Treatise, 78.

171   "and the yellow leg'd"; "stale"; "Their smell": Simmons, 8.

171–172   Take a Chicken: Crump, 140.

172–173   "Soup of Any Kind"; Put the fowls: Randolph, 18–19.

173   When the Fowl is plucked: Dixon, A Treatise, 77.

174   Put on the fire: Tyree, 79.

174   In order to make: Toulouse-Lautrec, 110–111.

175   Fricassee de Poulet: Toulouse-Lautrec, 111.

175   Chicken Marengo: Toulouse-Lautrec, 115.

182–183   "Place fresh eggs in cold water": Maestro Martino, 98.

183   Rumble Eggs: Tyree, 236.

183   Place fresh eggs: Lind, 336.

184   Eggs à la Crème: Tyree, 237.

Hen Music

203   Chickens are beautiful: Kilarski, 21.

204–205   Chickens are not pets: Smith and Daniel, 323.

Bibliography

Aesop. *Fables.* [Internet] Available from: http://tomsdomain.com/aesop/is181.htm.

Aldrovandi, Ulisse. *Ornithologiae.* [Internet] Available from: http://gdz.sub.uni-goettingen. de/no_cache/dms/load/toc/?IDDOC=234603.

Alper, Joe. "Hatching the Golden Egg: A New Way to Make Drugs." *Science*, Vol. 300, No. 5620 (May 2, 2003), pp. 729–730.

American Poultry Association. *American Standard of Perfection, 2010, Illustrated.* Burgettstown, Pennsylvania: American Poultry Association, Inc., 2010.

American School of Classical Studies at Athens. *Birds of the Athenian Agora.* Princeton, New Jersey: Institute for Advanced Study, 1985.

Aristotle. *History of Animals.* Translated by D'Arcy Wentworth Thompson. [Internet] Available from: http://classics.mit.edu/Browse/browse-Aristotle.html.

ARKive.org. Images and videos of red jungle fowl. [Internet] Available from: www.arkive. org/species/GES/birds/Gallus_gallus.

Burnham, George. *The History of the Hen Fever.* Boston, Massachusetts: James French and Company, 1855.

Burt, D. W. "Emergence of the Chicken as a Model Organism: Implications for Agriculture and Biology," *Poultry Science* 86 (2007), pp. 1460–1471.

Chaucer, Geoffrey. *Canterbury Tales.* Rendered into modern English by J. D. Nicholson. New York, New York: Garden City Publishing Company, Inc., 1934.

Columella. *On Agriculture*, Vol. II. Edited and translated by E. S. Forster and Edward H. Heffner. Cambridge, Massachusetts: Harvard University Press, 1968.

Crump, Nancy Carter. *Hearthside Cooking: Early American Southern Cuisine Updated for Today's Hearth and Cookstove*, 2nd edition. Chapel Hill, North Carolina: The University of North Carolina Press, 2008.

Darwin, Charles. *The Variation of Plants and Animals Under Domestication*. London, England: John Murray, 1868. [Internet] Available from: www.darwin-online.org.uk.

Dixon, Edmund Saul. *Ornamental and Domestic Fowl: Their History and Management*. London, England: Published at the Office of the "Gardeners' Chronicle," 1848. Download available from Google Books.

Dixon, Edmund Saul. *A Treatise on the History and Management of Ornamental and Domestic Poultry*, 3rd ed., with large additions by J. J. Kerr, M.D.Philadelphia, Pennsylvania: E. H. Butler & Co., 1853.

Ekarius, Carol. *Storey's Illustrated Guide to Poultry Breeds*. North Adams, Massachusetts: Storey Publishing, 2007.

Glasse, Hannah. *The Art of Cookery Made Plain and Easy: Which far exceeds any Thing of the Kind yet published.* A Google book: Enter the title in the search engine.

Heraclitus. *Herakleitos and Diogenes*. Translated by Guy Davenport. San Francisco: Grey Fox Press, 1979.

Hobson, Jeremy, and Celia Lewis. *Keeping Chickens: Getting the Best from Your Chickens*. Cincinnati, Ohio: David & Charles, 2010.

How Stuff Works. [Internet] Available from: http://science.howstuffworks.com/environmental/life/genetic/question85.htm.

International Chicken Genome Sequencing Consortium. "Sequence and Comparative Analysis of the Chicken Genome Provide Unique Perspectives on Vertebrate Evolution," *Nature*, Vol. 432 (December 9, 2004), pp. 695–716.

Kannampilly, Ammu. "Scientists Find Chickens Retain Ancient Ability to Grow Teeth." [Internet] Available from: http://abcnews.go.com/Technology/story?id=1666805.

Kilarski, Barbara. *Keep Chickens: Tending Small Flocks in Cities, Suburbs, and Other Small Spaces*. North Adams, Massachusetts: Storey Publishing, 2003.

Lazenby, Francis D., "Greek and Roman Household Pets." *The Classical Journal*, Vol. 44, No. 4 (Jan. 1947), 245–252, and Vol. 44, No. 4 (Feb. 1947), 299–307.

Lillico, Simon G., Michael J. McGrew, Adrian Sherman, and Helen M. Sang. "Transgenic chickens as bioreactor for protein-based drugs," *Drug Discovery Today*, Vol. 10, Issue 3 (February 1, 2005), pp. 191–196.

Lind, L. R., translator and editor. *Aldrovandi on Chickens*. Norman, Oklahoma: University of Oklahoma Press, 1963.

MacDonald, Betty. *The Egg and I*. New York, New York: Harper, 1945.

Maestro Martino di Como. *The Art of Cooking: The First Modern Cookery Book*, Luigi Ballerini, Editing and Introduction. Jeremy Parzen, Translation and Annotations. Berkeley, Los Angeles, London: University of California Press, 2005.

Moses, Anna Mary, and Otto Kallir. *Grandma Moses: My Life's History*. New York, New York: HarperCollins, 1952.

National Institutes of Health. "Trans-NIH Gallus Initiative." [Internet] Available from www.nih.gov/science/models/gallus.

Opie, Iona and Peter, eds. *The Oxford Dictionary of Nursery Rhymes*. Oxford, New York: Oxford University Press, 1966.

Pindar. *The Odes of Pindar: Including the Principal Fragments*. Translated by Sir John Sandys. Cambridge, Massachusetts: Harvard University Press, 1968.

Plato. *Plato in Twelve Volumes*, Vol. I. Cambridge, Massachusetts: Harvard University Press, 1966.

Pliny. *Natural History*, Vol. III. Translated by H. Rackham. Cambridge, Massachusetts: Harvard University Press, 1967.

Porta, John Baptista (Giambattista della Porta). *Natural Magick in XX Bookes*. Sioux Falls, South Dakota: NuVision Publications, 2005.

Randolph, Mary. *The Virginia Housewife Or, Methodical Cook*. A facsimile of an authentic early American cookbook. Reprint. Originally published: Philadelphia, Pennsylvania: E. H. Butler, 1860. New York, New York: Dover Publications, Inc., 1993.

St. John, James Augustus. *Egypt and Mohammed Ali, or, Travels in the Valley of the Nile*. London, England: Longman, Reese, Orme, Brown, Green & Longman, 1834. Digitized by Google.

Salatin, Joel. *Pastured Poultry Profits: Net $25,000 in 6 months on 20 acres*. Swoope, Virginia: 1993.

Schmutz, Jeremy, and Jane Grimwood. "Genomics: Fowl Sequenced," *Nature*, Vol. 432 (December 9, 2004), pp. 679–680.

Simmons, Amelia. *The First American Cookbook*. A Facsimile of *American Cookery*, 1796. New York, New York: Reprint. Originally published: *American Cookery*: New York, New York: Oxford University Press, 1958. New York, New York: Dover Publications, Inc., 1984.

Slavin, Philip. "Chicken Husbandry in Late-Medieval Eastern England: c. 1250–1400." *Anthropozoologica*, 2009, pp. 35–56.

Smith, Page, and Charles Daniel. *The Chicken Book*. Athens, Georgia: The University of Georgia Press, 1975.

Sunset magazine, "The chicken coop is the new doghouse," February 2011, page 35.

Swift, Jonathan. *Gulliver's Travels and Other Writings*. New York, New York: Bantam Books, 1962.

Tegetmeier, William Bernhard. *The Poultry Book: Comprising the Breeding and Management of Profitable and Ornamental Poultry*. London, England: George Routledge and Sons, 1867.

Theognis. *Elegy and Iambics*, Vol. I, J. M. Edmonds, ed. Available from: The Perseus Digital Library: www.perseus.tufts.edu/hopper. Enter "J. M. Edmonds" in the Search box.

Thompson, D'Arcy Wentworth. *A Glossary of Greek Birds*. London, England: Oxford University Press, 1936.

de Toulouse-Lautrec, Henri, and Maurice Joyant. *The Art of Cuisine*. New York, New York: Crescent Books, 1967.

Tsudzuki, Masaoki. "Japanese Native Chickens." Available from: http://www.scribd.com/tiniente/d/17056801-Japanese-Chicken.

Tyree, Marion Cabell. *Housekeeping in Old Virginia*. Richmond, Virginia: J. W. Randolph & English, 1878. Also available as a Google book. Enter the title in the search engine.

U.S. Food and Drug Administration. *Bad Bug Book,* 1st Edition. [Internet] Type Bad Bug Book in the search engine.

Varro. *On Agriculture*. Translated by William Davis Hooper, with a revision by Harrison Boyd Ash. Cambridge, Massachusetts: Harvard University Press, 1935.

Winstead, Edward R. "Meaningless Sex." Genome News Network. In the search box, type: Meaningless Sex – Genome News Network.

Wycliffe, John. English translation of the Bible. [Internet] available from: http://wesley.nnu. edu. Enter Wycliffe Bible in the search engine.

Zimmer, Carl. "The Evolution of Feathers," *National Geographic*, Vol. 219, No. 2 (February 2011), pp. 32–57.